The International Society for Gender Medicine

The International Society for Gender Medicine

History and Highlights

Edited by

Marianne J. Legato, M.D., Ph.D. (hon. c.), F.A.C.P.
Columbia University College of Medicine, New York, NY, United States
Johns Hopkins University, Baltimore, MD, United States

Marek Glezerman, M.D.
Rabin Medical Center, Petah Tikva, Israel

Academic Press is an imprint of Elsevier
125 London Wall, London EC2Y 5AS, United Kingdom
525 B Street, Suite 1800, San Diego, CA 92101-4495, United States
50 Hampshire Street, 5th Floor, Cambridge, MA 02139, United States
The Boulevard, Langford Lane, Kidlington, Oxford OX5 1GB, United Kingdom

Notices
Knowledge and best practice in this field are constantly changing. As new research and experience broaden our understanding, changes in research methods, professional practices, or medical treatment may become necessary.

Practitioners and researchers must always rely on their own experience and knowledge in evaluating and using any information, methods, compounds, or experiments described herein. In using such information or methods they should be mindful of their own safety and the safety of others, including parties for whom they have a professional responsibility.

To the fullest extent of the law, neither the Publisher nor the authors, contributors, or editors, assume any liability for any injury and/or damage to persons or property as a matter of products liability, negligence or otherwise, or from any use or operation of any methods, products, instructions, or ideas contained in the material herein.

British Library Cataloguing-in-Publication Data
A catalogue record for this book is available from the British Library

Library of Congress Cataloging-in-Publication Data
A catalog record for this book is available from the Library of Congress

ISBN: 978-0-12-811850-4

For Information on all Academic Press publications
visit our website at https://www.elsevier.com/books-and-journals

Publisher: Mica Haley
Acquisition Editor: Tari K. Broderick
Editorial Project Manager: Tracy Tufaga
Production Project Manager: Mohanapriyan Rajendran
Cover Designer: Mark Rogers

Typeset by MPS Limited, Chennai, India

CONTENTS

List of Contributors ix
Foreword xi
Preface xiii
Acknowledgments xvii

Chapter 1 Gender-Specific Medicine: An Idea That Should Have Been Intuitive But Which Required the Efforts of an International Community to Establish 1
Marianne J. Legato
References 8

Chapter 2 The Making of the Israel Society for Gender Medicine (IsraGem) 9
Marek Glezerman
References 15

Chapter 3 A View of the History of Sex/Gender Medicine in the United States 17
Vivian W. Pinn

Chapter 4 Gender Medicine in Austria 23
Alexandra Kautzky-Willer
References 31

Chapter 5 Gender Medicine in Cardiovascular Diseases: Past, Present, and Future 33
Jeanette Strametz-Juranek, Martin Skoumal, Sandra Steinböck and Angelika Hoffer-Pober
Introduction 33
Institutional Background—Gender Mainstreaming and Diversity at the Medical University of Vienna 34

Gender-Sensitive Research and Teaching 35
Gender Medicine in Cardiovascular Diseases: The Past 36
Gender Medicine in Cardiovascular Diseases: The Present 38
Gender Medicine in Cardiovascular Diseases: The Future 39
References 42

Chapter 6 Sex and Gender in Health: The World Writes on the Body 45
Gillian Einstein
Introduction 45
A Journey 46
The World Writes on the Body 48
Struggles and Support 50
Call for Action 52
Conclusion 53
References 53

Chapter 7 The Charité Approach—Sex and Gender in Research, Teaching, and Policies 57
Vera Regitz-Zagrosek
From a Student to Professor in Gender Medicine 57
Gender in Research 59
Teaching 60
References 63

Chapter 8 Gender-Specific Medicine in Italy: Point of View and Journey of Giovannella Baggio 65
Giovannella Baggio
Introduction: Personal Itinerary 65
Communication Tools Within and Outside the Network 76
Conclusion 77
References 78

Chapter 9 The Future of Science Through the Lens of Gender-Specific Medicine: Rewriting the Contract 79
Giuseppe Caracciolo and Alessandro Casini
References 82

Chapter 10 Overcoming the Skepticism to Reach Gender Equity and Appropriateness in Pharmacological Response 85

Flavia Franconi

Italian General Contest 85
Medical Contest..........86
Because I Studied Gender Pharmacology..........87
References..........91

Chapter 11 Gender Medicine in Italy: The Point of View of Maria Grazia Modena 95

Maria G. Modena and Valentina Martinotti

The Story..........95
Barriers..........101
The Last, Personal, Barrier: My Story as an Example of "Gender Related" Storm..........101
Limits and Plans for the Future..........104
References..........105

Chapter 12 Center for Gender Medicine Karolinska Institutet (KI) Since 2001 107

Karin Schenck-Gustafsson

How It Started..........107
Creating a Center for Research and Education..........110
National and International Activities..........112
Research..........113
References..........114

Chapter 13 From Appalachia to West Texas: A Journey to Move Beyond One-Sex Medicine 117

Marjorie R. Jenkins

Chapter 14 Sex and Gender in Emergency Medicine: Advancing Care Through Person-Specific Research, Education, and Advocacy 125

Alyson J. McGregor

Sex and Gender in Emergency Medicine..........125
Acknowledgment..........131
References..........131

Chapter 15 A Comparative Physiologist's Journey Into the Field of Sex- and Gender-Specific Medicine 133
Virginia M. Miller
Introduction 133
Trail Blazers 135
Progress in Spite of Rocky Roads 136
Not There Yet 138
Full Circle? 138
Acknowledgments 139
References 139

Chapter 16 A Short History of the International Society for Gender Medicine (IGM) 143
Marek Glezerman and Vera Regitz-Zagrosek
References 148

Index 149

LIST OF CONTRIBUTORS

Giovannella Baggio
University of Padua, Padova, Italy

Giuseppe Caracciolo
Menarini International Foundation, Florence, Italy

Alessandro Casini
Menarini International Foundation, Florence, Italy

Gillian Einstein
Department of Psychology, University of Toronto, Toronto, ON, Canada

Flavia Franconi
University of Sassari, Sassari, Italy; National Institute of Biostructures and Biosystems, Sassari, Italy

Marek Glezerman
Rabin Medical Center, Research Center for Gender and Sex-Specific Medicine, Petah Tikva, Israel; Tel Aviv University, Sackler Medical School, Tel Aviv-Yafo, Israel

Angelika Hoffer-Pober
Medical University of Vienna, Vienna, Austria

Marjorie R. Jenkins
Texas Tech University Health Sciences Center, Lubbock, TX, United States

Alexandra Kautzky-Willer
Medical University of Vienna, Vienna, Austria

Marianne J. Legato
Columbia University, New York, NY, United States; Johns Hopkins University, Baltimore, MD, United States

Valentina Martinotti
University of Modena and Reggio Emilia, Modena MO, Italy

Alyson J. McGregor
Brown University, Providence, RI, United States

Virginia M. Miller
Mayo Clinic, Rochester, MN, United States

Maria G. Modena
University of Modena and Reggio Emilia, Modena MO, Italy

Vivian W. Pinn
Former Director (Retired), NIH Office of Research on Women's Health, Bethesda, MD, United States

Vera Regitz-Zagrosek
Institute for Gender in Medicine and Center for Cardiovascular Research, Charité, University Medicine Berlin DZHK, partner site Berlin

Karin Schenck-Gustafsson
Center for Gender Medicine, Department of Medicine, Karolinska Institutet and Karolinska University Hospital, Solna, Stockholm, Sweden

Martin Skoumal
Pensionsversicherungsanstalt, Graz, Austria

Sandra Steinböck
Medical University of Vienna, Vienna, Austria

Jeanette Strametz-Juranek
Pensionsversicherungsanstalt, Graz, Austria

FOREWORD

The International Society for Gender Medicine: History and Highlights

The life work of academic visionaries like Dr. Marianne J. Legato, Dr. Marek Glezerman, and their colleagues from the International Society for Gender Medicine (IGM), has truly created new frontiers. Their innovative message has already echoed throughout the global scene and has paved the way for a new trend in modern medicine.

This book beautifully reflects unique aspects of the history of Gender and Sex Specific Medicine (GSSM) and its rise over time, outlining how this rapidly expanding field of science has shaped the horizon of tailored diagnosis and management of disease, considering the fundamental differences in gender/sex—and boldly disrupting the "*One Gender/One Sex medicine*" paradigm that has been mistakenly in place for years.

The personal experiences of these GSSM pioneers from seven countries serve as a catalyst for a much-needed multidisciplinary integration to objectively address the complexities surrounding GSSM, underscoring the need to re-write many chapters in medicine as we know it. It is in this spirit that a new congruent movement, the *Global Men's Health Foundation*, has emerged to promote global equality in access to health care—for both men and women—through continuous education, research, and effective legislations; fostering the interparticipation of academia, healthcare providers, medical/scientific associations, industry, insurance companies, patients, communities, and governments, in the process.

Notwithstanding the large body of evidence on the social and biological characteristics surrounding the topic, the lack of a comprehensive action plan has delayed the development and implementation of core elements and targets to realize the full potential of GSSM. With the explosion of GSSM, the field is at a turning point that requires accelerating the pace of this plan. It is important in ways that we have not yet begun to imagine.

Ingrid M. Perscky, M.D.
President and Founder, Global Men's Health Foundation

PREFACE

These sixteen stories of how each of us contributed to the revolutionary and indispensable discipline of gender-specific medicine (GSM) have been unexpectedly delightful to read. Some themes are common to all: the initial skepticism and lack of interest in gender medicine as an evidence-based, utterly essential concept was experienced by every one of us. All revolutionary ideas are met with resistance and incredulity and it seems to me that the bigger they are, the harder it is to ensure their acceptance, much less their implementation! Our first conferences were underfunded and our lectures were delivered in nearly empty rooms. Gradually and with enormous effort, the seats began to fill—but only with women; at first, nurses and female doctors were the only attendees.

Establishing GSM has been a labor of love, propelled by persuasion, perseverance, intelligence, and charm to weave our combined efforts into a universally accepted idea; that the normal function of men and women and their experiences of the same diseases are significantly different—so much so that understanding and implementing those differences in patient care would save lives. Remarkably, the assumption that all data harvested from studying males could be applied to females without separate verification characterized all biomedical research until the early 1990s. It was then that it first began to be apparent that women were not only small replicae of men, but, as it has turned out, **the sexes are profoundly and abundantly different in completely unanticipated ways.**

For virtually every one of us, the beginning was under the banner of improving our knowledge of how to care for women. We realized that we knew almost nothing about them beyond their reproductive biology and that there was a huge game of "catch up" to play before women's health evolved into GSM. This capitalized on the lay public's push in the last decades of the 20th century to improve the care of the female patient, and it was only gradually that we convinced our colleagues and society in general that **what we learned about women should**

be integrated into a comparison of the differences between the sexes. This expanded view has borne invaluable fruit.

Each of us learned that inviting collaboration and minimizing the sense of competition that invariably arises when a new approach is born are very important. As I read each of these narratives, the unique personality of each of the authors who accomplished that made his or her stories particularly fascinating. The paths to leadership roles in specific environments were varied: in countries where government support and control of health care delivery is extensive, regime changes made progress particularly difficult, as Maria Grazia Modena points out. Giovannella Baggio and Flavia Franconi recount the challenges to uniting many fragmented organizations under a single umbrella to pursue improvements in investigation and health care delivery. Some paths to gender-based medicine took unexpected turns: Virginia Miller's adventures with sleeping seals in a pool built in the middle of a desert were particularly engaging. The narratives from Austrians Jeanette Strametz-Juranek and Alexandra Kautzky-Willer illustrate their twin efforts to initiate and expand the science of GSM at the University of Vienna and then to implement its precepts into a unique, state-of-the-art gender-specific rehabilitation system for men and women with cardiovascular disease. Vivian Pinn's chapter resonated in a special way for me because of my close association with her as a member of her advisory council when she was building the concept of gender-specific science at the National Institutes of Health. Karin Schenck-Gustafsson brought gender-based medicine to the powerful Karonlinska Intitutet; her particular genius was mobilizing the support not only of the talented scholars there but combining it with the prestigious support of Sweden's Queen Sylvia. The Queen's eloquence, superb intellect, and charisma were key in augmenting public interest in GSM. Marjorie Jenkins showed a similar talent for enlisting powerful advocates when she persuaded Laura Bush to lend her name and efforts to the GSM center Marjorie established.

Alyson McGregor brought the idea of considering the sex of the patient into the emergency room for the first time; I can still remember the avidity with which an audience Alyson had organized heard the details of my own adventures in trying to break new ground for this novel idea.

In his chapter, Giuseppe Caracciolo tells one of the most heartening stories and is probably the newest convert to the importance of GSM; he explains how he is fashioning the Menarini Foundation's rich palette of conferences to incorporate the latest perspectives of the still young science of male-female differences.

Unquestionably the efforts of Vera Regitz-Zagrosek to introduce the principles of GSM into medical school curricula and training programs have been one of the most spectacular of all our achievements. She was aware from the beginning that GSM was not a "boutique" topic that deserved only to be a subset of the many individual disciplines of medicine. Rather, she saturated all the curricula at the Charité with the comparison between the physiology of the sexes and the differences in their experience of illness. Her work is a spectacular model for how we should all be refashioning our curriculum. Writing this, I am reminded of the time in 2000 I handed the first edition of the *Principles of Gender-Specific Medicine* to the head of the Columbia curriculum committee. He took the two volumes from me, looked at me skeptically and said: "Yes, but, how will we find somebody to teach this?"

As we navigate the genomic era, a time of unparalleled, even mind-blowing-new powers and ideas, Gillian Einstein sets out one of the most important issues we can now and must consider next: how the environment interacts with biological sex to produce the phenotype. As she puts it, "The world writes on the body." Precisely how that happens is one of the most interesting phenomena on our investigative agenda. The answers will resolve the old, outdated notion that sex and the environment are separate forces in fashioning the finished person. In fact, as we are more and more able to comprehend, they operate together in a final common pathway.

Marek Glezerman was one of the most effective pioneers of this new science; his powerful position in Israeli medicine gave unique authority to his message that GSM was the wave of the future. I will always think of him as my first colleague in this adventure. Unlike any of the rest of us, he filled the seats of his first congress with over 400 of the most intellectually able and fascinated health care practitioners I ever had the pleasure of meeting. The Israel Society for Gender Medicine was and remains a leader in the field. It is to him that we owe much of the character and strength of the International Society for Gender Medicine. His chapter, written with Vera Regitz-Zagrosek, describes the effort that went into making the Society a reality.

I am glad Tari Broderick, Sr. Acquisitions Editor, invited us to write this book; I think it was a good idea. Marek and I tried to have each of the contributors tell a deeply personal story, illustrating that there are as many ways of introducing a new concept as there are people invested in it. My thanks to all of our authors; all have my admiration for the successes they have achieved. Although there are common themes we all have shared, what I most enjoyed was that no story is the same. As is the case with all of our efforts, each is stamped with our own unique personality and style.

Marianne J. Legato, MD, PhD (hon. C.), FACP

ACKNOWLEDGMENTS

The editors thank our publisher, Tari Broderick, who proposed that we do this interesting project, nursed it to a speedy and satisfactory conclusion and reassured us that the world at large would enjoy reading about our joint adventures.

We particularly want to cite the contribution of Wendy Dauber, the Executive Director of the Foundation for Gender-Specific Medicine. Her expertise, intelligence, good humor, and unflagging interest in our work made not only this project possible, but animate, inform, and amplify every program and new venture we undertake, large or small. We believe that her contribution to gender-specific medicine is equal to that of any one of us. She is truly irreplaceable.

CHAPTER 1

Gender-Specific Medicine: An Idea That Should Have Been Intuitive But Which Required the Efforts of an International Community to Establish

Marianne J. Legato[1,2]
[1]Columbia University, New York, NY, United States [2]Johns Hopkins University, Baltimore, MD, United States

Writing a concise and interesting summary of my life in academic medicine and in particular about my role in establishing and expanding the new field of gender-specific medicine is not easy. I have read the other chapters in this book from the colleagues Dr. Glezerman and I have invited to tell their stories: each narrative is fascinating, unique, and deeply personal. I know most of these scholars quite well; I know about their struggles, triumphs, and disappointments; and I watched them persevere, blossom, and create bodies of work that they never even thought possible when they began their interest in gender-specific medicine. My collaboration with some were particularly rewarding, as evidence by my honorary memberships in the Israel, Austrian, and Korean Societies of Gender-Specific Medicine.

One of the first misapprehensions our community had to correct was that gender-specific medicine is not the study of women's health: it is the science of how the normal physiology and experience of disease differs between males and females at all levels of existing life. It took almost a decade for our colleagues to understand that we were not politically motivated feminists with a lopsided interest in women receiving more medical attention than they had heretofore enjoyed and that in fact, we were interested in the biology that defined men as different from women in all the systems of the body. In fact, the new knowledge about female biology often caused us to reframe and/or reinterpret what we knew about men. In other words, women were making an offer men could not afford to refuse.

The International Society for Gender Medicine. DOI: http://dx.doi.org/10.1016/B978-0-12-811850-4.00001-6

As I prepared to compose my own chapter, I began by printing out my curriculum vitae, which is 62 single spaced pages long—the accumulation of the work of four decades in academic medicine. Picking out the most meaningful landmarks was fun; the stories behind some of them are fascinating and worth telling about in some detail. I hope my review will be helpful to people entering this fascinating field and most of all show how life—even when it is meticulously planned—is full of unexpected twists and turns that make for a remarkable adventure.

I began life as an academically trained cardiologist, supported by the Martha Lyon Slater Fellowship and then a Senior Investigatorship from the New York Heart Association. The NIH then took over: my son was born on the day the NIH notified me that I had won a Research Career Development Award, prompting a note from my chairman remarking on the excessive nature of the double achievement. The NIH continued to support my investigative work on the ultrastructure and function of the myocardial cell. I served on NIH cardiovascular study sections and had an unparalleled opportunity to learn about the work and expertise of the leaders in my field.

These first years of my training and research were set in a period when the whole biomedical community studied only males at all levels of investigation and for economy's sake, often used animal models, suggesting that the data told us at least something about human physiology. Looking back on this period, I referred to it as characterized by "a bikini view" of women; we assumed that only their breast and gynecologic health were unique, and that in all other respects, they were physiologically interchangeable with men. It followed, then, that all human's experience of disease was identical. What an inexplicable (and inexcusable) intellectual error and what a long way we have come!

My own particular epiphany came in 1992 with a visit from a journalist, Carol Colman, whose mother had died of coronary artery disease (CAD). At the suggestion of the American Heart Association, she asked me to collaborate with her on a book about women's experience of this illness; she was convinced that her mother's risk factors, course, and outcome from CAD were not only different from that of men, but that the specific features of her illness were totally overlooked by the physicians who treated her. We wrote "The Female Heart: The Truth About Women and Coronary Artery Disease" [1], and in 1992 the

American Heart Association awarded us the Blakeslee Award for the best book about heart disease for the lay public that had been published that year.

The work on the book was a paradigm shift for me, and it was the first step in my entry into gender-specific medicine. I had no idea about whether there was a difference in women's experience of CAD and that ignorance of those differences was costing women their lives. So many things were wrong about the way we viewed women with heart disease: we believed that they were relatively immune from CAD until old age, that their distinctive symptoms were often the consequence of anxiety and other neuroses, and our treatment of them was informed by the conviction that their frailty precluded them from the kind of aggressive treatment offered to men.

In an amazing turn of events, I received an invitation to have tea with one of the men who had served as a judge for the American Heart Blakeslee award, T. George Harris, founder of *Psychology Today*. At that time, he was a consultant in women's health for Procter and Gamble. He asked me if I wanted to become an advisor to the company about products created specifically for women. I replied that he had no way of knowing whether I would be a valuable consultant for a major American company, but that I had a better idea: I proposed a partnership between Procter and Gamble and Columbia University as a terrific way of opening the company's access to expertise of all kinds.

Mr. Harris liked the idea, and I began a 2-year effort to propose the collaboration and make it a reality. I could never have done it without the support of my chairman, Dr. Myron Weisfeldt, whom I approached with the idea that (1) if women's hearts were so different from men's hearts, I believed other organs might be different as well and (2) that a union with a huge American corporation would be a boon for Columbia scholars. Dr. Weisfeldt's support for these ideas never wavered—although most of the rest of the faculty thought it was a wild and doomed project and said so.

Two years of intensive work began; I used everything I could to convince P&G that we were worth an investment. Because there was nothing in the literature summarizing differences between men and women, I went to the library with my secretary (this was well before the years of Google Scholar and the internet) and we combined the library stacks at Columbia for material to produce a book, *Gender-Specific Medicine for the Practicing Physician* [2]. I knew I would need that to prove to P&G's corporate chiefs and their research community that gender-specific biology was a fact and that products could be made based on our understanding of the unique needs of both sexes.

I met everyone of note in the P&G executive office during those 2 years. Two men stand out in particular: Craig Wynett, still there as a special advisor to the chairman and John Pepper, then chairman and CEO of P&G. P&G's vetting of me was thorough: on one occasion, I addressed their 1500 researchers, most of them PhD's, on the differences between the physiology of men and women. I remember the day my 2 years of work on the alliance were almost finished: John Pepper invited me to a private lunch in his office and opened with an unforgettable question: "Do you have the confidence of your university?" I assured him I had, although my university support at that moment consisted of Dr. Weisfeldt, period.

Finally, P&G's top executives came to Columbia for a visit with Dean Herbert Pardes (who clearly regarded me with a great deal of skepticism). Mr. Pepper addressed the Dean with the following remark: "We have a real interest in supporting a liaison with Columbia as a result of our work with Dr. Legato, and if our discussions continue to be fruitful for another year we will support an alliance." I rose to my feet: "Mr. Pepper, I have been in negotiations with you for two years. I have nothing more to explore, produce or offer, and if you don't support us on the basis of what we have been able to achieve together at this point, I am giving up the idea." The room was full of shocked silence. Fortunately, Mr. Pepper burst out laughing and a week later, Craig Wynett gave my chairman a check for a million dollars to start the Partnership for Women's Health, later to become at my incessant urging The Partnership for Gender-Specific Medicine. One of the most telling moments in that accomplishment was Dr. Weisfeldt's remark as I handed him the check: "Now we have to start a search for a director for the new program." I was incredulous. I reminded him that I had worked for 2 years on the idea, developed it, achieved this improbable union with P&G, and I fully expected to be its director. He immediately agreed, fortunately, and we were off to build the program together.

A second memorable outcome of the negotiations was that I was called before two of our senior vice presidents in the course of finalizing our agreement with P&G, who questioned me about whether or not I had taken any money from P&G to work on this negotiation. The idea of taking money from P&G to achieve a partnership with Columbia had never occurred to me! To his everlasting credit, Dr. Weisfeldt, who was conducting a symposium in Florida at the time, flew back to be with me at this unexpected hearing, saying: "If they're going to attack Legato they'll have to attack me too!" I will never forget his defense of my integrity at that encounter (or the appalling fact that it was not necessary at all for him to have to defend me). Apparently my examiners were satisfied, but as a parting shot one of the two said to me: "Just don't take any stock options from P&G." Clearly, this kind of a negotiation (to engineer a union between one of America's most powerful corporations and the medical school) was unique in their experience—as it was for me.

I tell this story in this much detail because I know that many investigators have set themselves wildly improbable goals and met with skepticism and setbacks in their attempt to achieve them. Other chapters in this book tell other stories like this one. The importance of having a mentor like Dr. Weisfeldt is an essential ingredient for success. So is the ability to think of how to exploit every opportunity for accomplishment in the byzantine world of academic medicine.

In any case, the 4-year award of 4 million dollars from P&G enabled us to produce two indexed journals and two editions of the first major textbook on the new science (*Principles of Gender-Specific Medicine*) in the years between 1998 and 2011. We also raised a million-dollar fund for a named professorship in Gender-Specific Medicine; we still use the income from that fund to support junior Columbia faculty: The M. Irené Ferrer Professorship in Gender-Specific Medicine, a title currently held by Dr. Elaine Y. Wan, Assistant Professor of Medicine at Columbia.

Because the chairman who followed Dr. Weisfeldt did not have any interest in the program I had begun, I resigned my full-time position at Columbia and transformed the Partnership for Gender-Specific Medicine into a private foundation, the Foundation for Gender-Specific Medicine. We have supported and continue to support scholars (two each year for the last two years at Johns Hopkins with a commitment for two more at that institution this year and six at Columbia), conduct symposia, and have produced a third edition of *The Principles of Gender-Specific Medicine*, with contracts from Elsevier for this book and for my next book on variations in human sexuality. With Mary Ann Liebert, Inc., publishers, I founded my third journal, *Gender and the Genome*, which has just been born and whose second issue is now in print.

Another important body of my work in gender-specific medicine was accomplished in a very fruitful collaboration with Dr. Vivian Pinn at the Office of Research on Women's Health at the NIH. She was a brilliant and fascinating colleague. I served as a charter member of her advisory committee from 1995 through 1998 and had the remarkable opportunity of co-chairing a series of four nation-wide symposia about the state of women's health. The result was our final report, *Beyond Hunt Valley: Research on Women's Health for the 21st Century*. I loved that time at the NIH with Dr. Pinn: her advisory committee was a hotbed of arguments about whether biomedical scientists were consummate snobs who held the sociologic community in contempt and had no interest in separating the contribution of the environment to phenotype. What is biological sex and what is the consequence of environmental experience is still debated, but it is clear not only that both are essential components of the phenotype but operate through a final common pathway to produce the unique individual. One of my chief goals, in fact, is to find a *single word* that will describe the dual contribution of biological sex and the environment ("gender"). I am going to ask the IGM to think about such a word at our Japan meeting in September of this year, and will offer "gensex" as a starting suggestion.

Most of the scholars in this book list a spectacular sequence of awards and citations, including honorary PhD's as a result of their achievements in the most recent years of their careers. I have some as well, the most significant of which to me are an honorary PhD from the University of Panama in 2015 and most recently an honorary award for Excellence in Science from the University of Messina in November of 2016.

My co-editor, Dr. Glezerman, urged me not to close my chapter without stressing the importance of cultivating and fostering the interest of laymen and women in the work we do. I have always believed that research dollars follow the interests of the public—not what scientists think is important. There was tremendous pushback at first to the idea that there was gender prejudice in health care and that women were suffering from our ignorance about the unique aspects of their basic physiology and their experience of disease. Shortly after Colman and I published our book on women's experience of heart disease, a producer from the television show, 20/20, asked me if I were willing to appear on the show and say that women were being discriminated against by the medical establishment. He confided that no cardiologist he had approached had been willing to say that publicly but I felt we had everything to gain by highlighting the issues female patients were facing. Agreeing to appear on 20/20 was my debut in the public arena as an advocate of gender-specific research and patient care.

I have accepted every invitation to talk to the public that I was privileged to be offered: my colleagues and I worked with luminaries like Oprah Winfrey, the leaders of Hadassah, Larry King and Mehmet Oz. But small venues were just as important; people throughout the United States and the whole world found the concept of gender-specific medicine fascinating. One of the most fruitful periods of collaboration I had was in Japan with a prominent journalist, Mitsuko Shimomura, who with her own resources founded a center for women's health in the middle of Tokyo that began the interest of that country in gender-specific medicine. My own patients have made financial and intellectual contributions to our efforts in New York that have enabled our Foundation to support research by young investigators, to establish three scientific journals and produce the three editions of our textbook on GSM [3]. These three volumes have chronicled the development of the science of GSM from its beginnings to the present era; the third edition has just appeared and describes GSM in the genomic era.

While our scientific publications were a natural product of our work, I felt it was equally important to write books for the lay public to nourish and encourage their support. The first of these was *Eve's Rib* [4] to summarize for laymen the scope and nature of the differences between men and women; Marek Glezerman steadfastly maintains that reading that book began his interest in gender-specific medicine. His invitation for me to come to Israel to help explore GSM in his first symposium in Tel Aviv in 2009 was one of the most exciting developments in establishing an international program for GSM. Those few days with Marek and his colleagues began a collaboration and friendship that was one of the most valuable ingredients in what became a sturdy and ever-expanding international project.

I have enjoyed my academic life immensely, as well as the care of my patients, many of whom have responded to my relationship with them by supporting my Foundation, making it possible for me to interface with both the professional and lay public in spreading the importance of the work of the last 30 years. Founding the International Society for Gender-Specific Medicine with other colleagues and pioneers (who describe their roles in their own chapters of this book) was a huge step forward in moving an interest and expertise in the science forward. The stories of how we all did it are inspiring; the collegiality of our community is unique and a rich source of collaborative effort. I am very grateful for their involvement in our mutual enterprise.

REFERENCES

[1] Legato MJ, Colman C. The female heart: the truth about women and coronary artery disease; 1991.

[2] Legato MJ. Gender specific aspects of human biology for the practicing physician. Armonk: Futura; 1997, 141 pp.

[3] Legato MJ. Principles of gender-specific medicine. Gender in the genomic era. London: Academic Press; 2004 (vol. I, 625 pp; vol. II, p. 627–1245; Second edition, 770 pp, 2010; Third edition, 2017).

[4] Legato MJ. Eve's Rib: the new science of gender-specific medicine and how it can save your life. New York: Harmony Books; 2002, 258 pp.

CHAPTER 2

The Making of the Israel Society for Gender Medicine (IsraGem)

Marek Glezerman[1,2]
[1]Rabin Medical Center, Research Center for Gender and Sex-Specific Medicine, Petah Tikva, Israel
[2]Tel Aviv University, Sackler Medical School, Tel Aviv-Yafo, Israel

My interest in Gender-Specific Medicine arose after I read "Eve's Rib" by Marianne Legato [1]. Chairing a busy department of Obstetrics and Gynecology with the associated commitments to clinical duties, research, teaching, and administration, reading time is very precious and needs to be allocated very carefully. But this book caught my eye and I read it as a break from so many "must reads." My first reaction was rather skeptical. How could it be that there were so many physiological and pathophysiological differences between the sexes when during almost four decades of a medical career, this had never consciously occurred to me? Of course, one looks for excuses and mine came quickly. My patients are all female, I told myself, so why bother about sex and gender differences? But then again, in reproductive endocrinology, which is a subspecialty in my medical discipline, we have female and male patients and in obstetrics there are female and male fetuses. So yes, if there are indeed sex and gender differences, I thought it would certainly make sense for me to reflect more on this issue. But what if all these differences were merely environmental, meaning nurture and not nature? Then of course, we would be talking about sociological, cultural, and environmental aspects with very different medical implications. The year was 2007–8 and the literature was not overabundant although quite a few high-quality studies had been published already. In the United States, almost 20 years earlier, Congress had passed the Women's Health Equity Act, which dedicated research funds to women's health, and the National Institute of Health had established the Office for Women's Health and large studies like the Women's health Initiative had been started. Yet, the general public and medical community were largely ignorant about the impact of

The International Society for Gender Medicine. DOI: http://dx.doi.org/10.1016/B978-0-12-811850-4.00002-8

biological sex and gender on clinical medicine. But rather surprisingly for me, there were even two textbooks available: One, a two volume-textbook edited by Prof. Marianne Legato, a cardiologist from Columbia University [2], and one edited by Prof. Anita Rieder, a medical sociologist from Vienna together with Prof. Brigitte Lohff, a historian and ethicists from Germany [3]. There were more to follow [4–6]. I also found a small paperback written over 30 years earlier by Michael Teitelbaum [7] but seemingly forgotten since. Intensive searching (Google was still not what it is today) led me to the conclusion that yes, there was quite some activity around the world but very little connection between those involved. In the United States, the newly founded OSSD (Organization for the Study of Sex Differences [8]) had begun its activities, but there was apparently no connection between the various groups.

So I arranged for myself a 3-month Sabbatical, aimed at mapping the existing activities in the field of Gender- and Sex-Specific Medicine (GSSM) and to visit a few centers. Everything was surprising for me: I learned that in leading academic institutions in Australia, Austria, Germany, Italy, Sweden, and the United States, research was being conducted and teaching programs were being developed. I learned about the existence of a newly founded International Society for Gender Medicine [9] which had already conducted two international meetings.

Back from my Sabbatical I presented a summary of what I had seen and learned to the board of department chairs at the Rabin Medical Center, where I was heading the Hospital for Women at that time. At the outset of my presentation, I asked my colleagues, all distinguished and accomplished clinicians, teachers, and scientists what they knew about Gender Medicine. Those few who volunteered information, interpreted Gender Medicine as a variation of Gender studies, that is mainstreaming women in the public health system and struggling for equality of women at the workplace, in teaching, research, promotion, and remuneration. Just as I had been utterly surprised at my first encounter with gender aspects of medicine, so were my distinguished colleagues. The meeting extended long beyond the planned time frame' there was a sense of electrification in the room and the discussion was heated. There were skepticisms and very valid questions were raised which I have encountered many times since. How come that these issues had never been addressed before? Was it really believable that most research was being performed almost exclusively in men? Why would we need gender-specific medicine, if we are on the verge of embarking on the path of personalized medicine? If we emphasize the difference in physiology and pathophysiology between men and women, would this not be just another attempt to cement the differences between the sexes and thus further support the longstanding discrimination of women? If we differentiate between women and men, would we then also need to differentiate between the adult and the geriatric population, between whites and blacks and many other subgroups? I have addressed some of these very valid issues elsewhere [10]. Toward the end of the meeting and spontaneously an interest group, comprising 21 department chairs from different medical disciplines, was created and went to work almost immediately. Many of us decided as a first step to look at our own data and to see whether a statistical re-run including gender stratification would be reasonable. The results were amazing: As expected, most of the previous clinical studies which included both sexes were not stratified according to sex. Thus re-examining data sets from various disciplines could commence almost immediately and the output was both rewarding and motivating to initiate prospective studies. Within a rather short period of time, other colleagues from other hospitals and universities joined and a few months later we decided that it was time for the foundation of an Israel Society for Gender Medicine. To my good fortune I discovered that the Third International Meeting on Gender Medicine was upcoming in 2008 in Stockholm, sponsored by Prof. Karin Schenck-Gustafsson, founder and head of the Center for Gender Medicine at the Karolinska Institute. Karin had practically introduced Gender Medicine in Sweden and had been able to recruit for this purpose the queen of Sweden. In Stockholm I met the women pioneers who formed the board of the International Society for Gender Medicine. They had already organized two international meetings in the past, in Berlin and Vienna. These were the women who were doing and had been doing so much to promote GSSM in their respective countries and internationally. They had practically created a new discipline and the names of all of them were familiar to me from my background research. These women were Prof. Marianne Legato, cardiologist from the United states, already a legend in her own time who had edited the first textbook on Gender Medicine, created the Foundation of GSSM at Columbia University and who was a founding editor of the first and at that time only journal dedicated to GSSM; Prof. Jeanette

Strametz-Juranek, cardiologist and founding president of the Austrian Society for Gender Medicine; Prof. Maria Grazia Modena from Italy, cardiologist and founder of the "Women's Clinic" in Modena and last but not least, Prof. Vera Regitz-Zagrosek, cardiologist, at that time president of IGM and the founding director of the Institute for Gender Medicine at the venerable Charite in Berlin. All were true pioneers and distinguished clinicians and researchers determined to introduce a new discipline into medicine. During this congress, elections were conducted and I was elected to the board. Now, I was really the odd-man out: I was the only male and also the only noncardiologist on the board. Nevertheless, I immediately felt very welcome. During the deliberations at this Stockholm meeting, I arrived at two understandings which proved to be of great significance to me: one was the strong perception that I was being giving the great privilege of participating in a historical turning point, namely the development of a novel viewpoint; a new angle from which to look at medicine. The situation was similar to that of some two centuries ago when physicians realized that children were not just small adults.Whatever was known related to physiology and pathophysiology of adults could not simply be applied to children by just correcting for weight and size. Subsequently the new medical discipline of Pediatrics was born. And now we were deliberating on what seemed to be quite a no-brainer which had been neglected for such a long time: Women are not just small men and therefore it was utterly inconceivable that more than 75% of research on health and disease was being conducted on men and extrapolated to women. Historical turning points are usually identified by hindsight, that is when looking back at history it becomes apparent that at this and that point in time something very meaningful happened which fundamentally changed the subsequent course of events. In Stockholm, I felt strongly and with awe that I was witnessing the creation of such a turning point in medicine. My second important understanding was that I should be part of this process and devote time as much as possible which I could chop out of my busy schedule to help and promote GSSM. So, the next steps would be to create a society for GSSM in Israel, to organize a scientific meeting, to start spreading the word in my country as loud and as much as I could, and to initiate research projects. The odds seemed to be good. At that time I held different offices, the most important of them, Director of the Women's Hospital at the Rabin Medical Center and deputy director of this probably largest tertiary health care center in Israel. I was also chair of Obstetrics and Gynecology at Tel Aviv

University. I confess that I was determined to use all the resources associated with these functions shamelessly in order to promote GSSM.

In order to give the planned congress the largest possible impact I intended to bring to Israel the internationally best-known proponent of GSSM at that time. And yes, Marianne Legato who has since become a very dear friend to me accepted the invitation and gave the keynote at our first Israel National congress on GSSM which was also the foundation meeting of our new society in February 2009. The meeting was a huge success attracting over 450 physicians, nurses, students, and professionals from many disciplines. At that meeting GSSM was introduced in Israel and is now established and thriving. The new board reflected and continues to reflect the multidisciplinary character of our new society. Currently the board includes two gynecologists (Prof. Dov Feldberg and myself), an internist (Prof. Ilan Krause), an internist and clinical pharmacologist (Dr. Heschi Rotmensch), a psychiatrist (Dr. Pnina Dorfman-Etrog), and a registered nurse (Ms. Orith Barak). We are conducting annual national congresses, each dedicated to a specific medical topic. These meetings have already become something like a tradition. They are all conducted at and are generously sponsored by the Rabin Medical Center. Although the meetings are aimed at health care professionals, the general public is welcome and we proudly host up to 200–250 participants at each meeting. So far the topics have covered sex and gender aspects of cardiology, nutrition, medications, psychiatry, obesity, and pain. We are organizing postgraduate courses at Tel Aviv University at the open university, our members are lecturing at all academic institutions in Israel and hospitals, and GSSM has been introduced officially into the curriculum of medical studies in two out of the five medical faculties in our country. We were very fortunate to be strongly supported by the previous dean of the Medical Faculty, Prof. Joseph Mekori and later by the current dean, Prof. Ehud Grossman and the head of the Medical School, Prof. Iris Barshak in our efforts to include GSSM in the teaching program of medical students. Today, GSSM is fully integrated in the teaching program at Tel Aviv University. Our website and more so, our facebook page is vibrant with tens of thousands visitors which is not bad for a webpage in Hebrew. The research center for Gender Medicine at the Rabin Medical Center, which I had to honor to create and which I am currently directing, has initiated and coordinated dozens of research projects in many medical disciplines (http://www.isragem.org.il/?CategoryID=275). My own research has been focused on sex aspects of intrauterine and perinatal sex differences. I published a widely read editorial in which I presented the new science of GSSM to the medical community in Israel [11]; we studied the influence of fetal sex on pregnancy outcome in normal singleton pregnancies [12] and in twin pregnancies [13]; we addressed the importance of GSSM in pediatrics [14], discussed the "feminization" of obstetrics and gynecology [15], developed sex-specific ultrasonographic methods in order to increase measurement accuracy [16,17], and performed the studies on sex differences in fetal growth [18,19]. We studied sex differences in birth traumas and stillbirth [20–22] and gender aspects of fetal programming [23,24]. We also addressed the ongoing debate about "the female and the male brain" [25,26]. Gender and Sex aspects in other medical disciplines were also included in my research [27].

Back to the year 2009: the Israel Society for Gender Medicine had just entered the world of Gender Medicine and with quite some Chuzpe, we offered to organize the next international congress on GSSM in Tel Aviv. The board accepted our invitation and in 2010 we proudly hosted the 5th International Congress. The keynote lectures were given by Nobel Laureate Prof. Aaron Ciechanover and by the honorary Congress President Prof. Marianne Legato. Without intention the two topics chosen by the lecturers reflected the ongoing discussion about the relation between Personalized Medicine and Gender Medicine. We had about 450 participants from 18 countries. At the opening ceremony, attended among others by the president of the Weizman institute, the president of the Israel Academy of Sciences, and the dean of the Medical School at Tel Aviv University, the director of Rabin Medical Center, Dr. Eyran Halpern, announced his decision that his Center would be the first in Israel dedicated to the promotion of GSSM. This declaration and promise has since materialized many times in the form of research grants, the establishment of a research center for Gender Medicine, the creation of a specialized Center for the treatment of women's hearts, led by Dr. Tal Porter and much more. In Israel, GSSM is here to stay; the topic has attracted many students for their MD theses; various professional groups organize now their own professional meetings dedicated to SGSM; and the topic has become very popular on talk shows, radio, and TV. We have presented at various parliamentary meetings and our voice is being heard clear and loud. So far for a brief history of the Israel Society for Gender Medicine (IsraGem, www.isragem.org.il). A historical change has taken place. It is now upon us not to let the momentum slow down.

REFERENCES

[1] Legato JM. Eve's Rib. New York: Harmony Books; 2002.

[2] Legato MJ, editor. Principles of gender-specific medicine. San Diego: Elsevier; 2004.

[3] Rieder A. gender medicine. In: Lohff B, editor. Geschlechtsspezifische Aspekte für die klinische Praxis. Wien, New York: Springer; 2005.

[4] Schenck-Gustafsson K, editor. Handbook of clinical gender medicine. Basel: Karger; 2012.

[5] Oertelt-Prigione S, Regitz-Zagrosek V, editors. Sex and gender aspects of clinical medicine. London: Springer Verlag; 2012.

[6] Kautzy-Willer A. Gendermedizin. Wein: Boehlau Verlag Wien; 2012.

[7] Teitelbaum SM. Sex differences. Garden City: Anchor Press/Doubleday; 1976.

[8] <http://www.ossdweb.org/>.

[9] <http://isogem.com/>.

[10] Glezerman M. Gender medicine. New York and London: Overlook/Duckworth; 2016.

[11] Glezerman M. Discrimination by good intention—gender based medicine. Isr Med Assoc J 2009;11:39–41.

[12] Melamed N, Yogev Y, Glezerman M. Fetal gender and pregnancy outcome. J Matern Fetal Neonatal Med 2010;23(4):338–44.

[13] Melamed N, Yogev Y, Glezerman M. The effect of fetal sex on pregnancy outcome in twin pregnancies? Obstet Gynecol 2009;114(5):1085–92.

[14] Glezerman M. For debate: is gender medicine important in pediatrics? Pediatr Endocrinol Rev 2009;6(4):454–6.

[15] Rabinerson D, Kaplan B, Glezerman M. [The feminization of obstetrics and gynecology]. Harefuah 2010;149(11): 729–32, Review. [in Hebrew].

[16] Melamed N, Yogev Y, Ben-Haroush A, Meizner I, Mashiach R, Glezerman M. Does the use of a sex-specific model improve the accuracy of sonographic weight estimation? Ultrasound Obstet Gynecol 2012;39(5):549–57.

[17] Melamed N, Ben-Haroush A, Meizner I, Mashiach R, Glezerman M, Yogev Y. Accuracy of sonographic weight estimation as a function of fetal sex. Ultrasound Obstet Gynecol 2011;38(1):67–73.

[18] Ben-Haroush A, Melamed N, Oron G, Meizner I, Fisch B, Glezerman M. Early first-trimester crown-rump length measurements in male and female singleton fetuses in IVF pregnancies. J Matern Fetal Neonatal Med 2012;25(12):2610–12.

[19] Melamed N, Meizner I, Mashiach R, Wiznitzer A, Glezerman M, Yogev Y. Fetal sex and intrauterine growth patterns. J Ultrasound Med 2013;32(1):35–43.

[20] Melamed N, Aviram A, Silver M, Peled Y, Wiznitzer A, Glezerman M, et al. Pregnancy course and outcome following blunt trauma. J Matern Fetal Neonatal Med 2012;25(9):1612–17.

[21] Hadar E, Melamed N, Sharon-Weiner M, Hazan S, Rabinerson D, Glezerman M, et al. The association between stillbirth and fetal gender. J Matern Fetal Neonatal Med 2012;25(2):158.

[22] Linder I, Melamed N, Kogan A, Merlob P, Yogev Y, Glezerman M. Gender and birth trauma in full-term infants. J Matern Fetal Neonatal Med 2012;25(9):1603–5.

[23] Glezerman M. Gender aspects of fetal programming. In: Schenk-Gustafsson K, deCola P, Pfaff D, Pisetsky D, editors. Handbook of clinical gender medicine. Basel: Karger; 2012. p. 37–50.

[24] Glezerman M. Intrauterine development of sex differences—fetal programming. In: Legato MJL, editor. Principles of gender specific medicine. 3rd ed. London: Liebert Elsevier; 2017.

[25] Glezerman M. Yes, there is a female and a male brain: Morphology versus functionality. Proc Natl Acad Sci USA 8 March 2016; pii: 201524418. PMID: 26957594.

[26] Glezerman M. The problem of the definition and quantification of reality. Ital J Gender-Specific Med 2016;2(1):3–4.

[27] Dickman R, Vainstein J, Glezerman M, Niv Y, Boaz M. Gender aspects suggestive of gastroparesis in patients with diabetes mellitus: a cross-sectional survey. BMC Gastroenterol 2014;14:34.

CHAPTER 3

A View of the History of Sex/Gender Medicine in the United States

Vivian W. Pinn
Former Director (Retired), NIH Office of Research on Women's Health, Bethesda, MD, United States

During the past 25 years, the scientific and medical communities have intensified efforts to define the broader context of women's health beyond traditional concepts, which had been primarily limited to the female reproductive system or so-called "Bikini Medicine" as it was aptly characterized by Dr. Marianne Legato. Now, expanded concepts inclusive of conditions that affect both women and men across the lifespan bring into focus the importance of determining sex and gender influences, based upon biomedical research that provides data to document such differences for application in an evidence-based medical care environment. From early advocacy demands to include women in clinical research protocols, to the current status of identifying the effects of sex and gender factors ranging from molecular and genomic science to the translational aspects of health and disease, women's health and health care have evolved into Gender-Specific Medicine (GSM), bringing a broad and comprehensive approach to medical care.

The historic lack of knowledge about gender differences related to cardiovascular disease was, and continues to be, a prime example of the stimulus to assess sex differences in etiology, prevention, diagnosis, treatment, and outcomes, but there are many and diverse areas of human development, health, and disease for which information about gender differences was lacking, and since have been or are now being discovered. There was initially reticence by some scientists who considered the initiatives to address women's health research only a reflection of being "politically correct." Fortunately, that has been corrected with

The International Society for Gender Medicine. DOI: http://dx.doi.org/10.1016/B978-0-12-811850-4.00003-X

advances in biomedical research, which have demonstrated the potential clinical significance of sex-specific mechanisms of pathophysiology, and the policies for inclusion were developed based upon scientifically appropriate principles.

It was a coalition of women's health advocates, scientists, and legislators who expressed concern that clinical research receiving federal funding by the National Institutes of Health (NIH) did not consistently include women, and if included, not in significant numbers to determine if sex and gender differences existed in the response to the intervention being studied. A landmark request from the Executive Committee of the Congressional Caucus for Women's Issues in 1990 for the meeting with the leadership of the NIH led to the formal initiation of the federal government's attention to women's health. This meeting, requested by the bi-partisan Co-Chairs of the Caucus, Representative Patricia Schroeder of Colorado, and Representative Olympia J. Snowe of Maine, and also Representative Constance Morella of Maryland and Senator Barbara Mikulski also of Maryland, was to determine the intent of the NIH to fully implement policies that would specifically determine sex and gender differences in health through the inclusion of women in clinical research. In response, when that meeting occurred on September 10, 1990, the NIH announced that it was establishing the Office of Research on Women's Health (ORWH) to ensure that women would be included in biomedical and behavioral clinical research studies, thus giving origin to the first office within the U.S. Department of Health and Human Services to have women's health as its primary mission and to the recognition of sex and gender as important variables in NIH funded research. The ORWH was also tasked with strengthening research regarding women's health by determining priorities for future investigation through a national agenda that identified gaps in knowledge. Other offices of women's health or positions responsible for women's health were subsequently established in the Office of the Secretary of HHS and within other agencies of the Department of HHS. The Congressional Caucus on Women's Issues further codified the NIH policy for inclusion of women and minorities by leading the Congress to include in the NIH Revitalization Act of 1993 (Public Law 103-43) a requirement that women be included in clinical studies in numbers such that a valid analysis of gender differences could be determined by the results.

EXECUTIVE COMMITTEE
Patricia Schroeder, Co-chair
Olympia Snowe, Co-chair
Lindy (Mrs. Hale) Boggs, Secretary
Marcy Kaptur, Treasurer
Barbara Boxer
Cardiss Collins
Nancy Johnson
Nancy Kassebaum
Barbara Kennelly
Jill Long
Nita Lowey
Jan Meyers
Barbara Mikulski
Susan Molinari
Constance Morella
Mary Rose Oakar
Elizabeth Patterson
Nancy Pelosi
Patricia Saiki
Claudine Schneider
Louise Slaughter
Jolene Unsoeld

Congressional Caucus
for
Women's Issues
Congress of the United States
Washington, D.C. 20515

2471 Rayburn Building
(202) 225-6740

Leslie Primmer
Executive Director

August 22, 1990

Dr. William Raub
Acting Director
National Institutes of Health
Bethesda, Maryland 20892

Dear Dr. Raub:

We are writing to propose a meeting between members of the Executive Committee of the Congressional Caucus for Women's Issues and key government health officials to discuss women's health research issues. We hope that this meeting can take place at the National Institutes of Health on September 10 at 10 a.m. In addition, we would like the public to be able to attend.

The purpose of the meeting would be to provide members of the Caucus with an opportunity to learn what steps have been taken by NIH to respond to issues raised in the GAO report and elsewhere, and to discuss with key government officials what steps may be necessary to strengthen federal research efforts on women's health issues. In addition to yourself, Dr. Ruth Kirschstein and others at NIH, we are inviting Secretary Sullivan and Surgeon General Novello to participate. Separate letters have been sent to each of them.

We believe that this meeting will provide an important opportunity for us to work together constructively to improve the health status of American women.

Sincerely,

Patricia Schroeder, Co-Chair
Member of Congress

Olympia J. Snowe, Co-Chair
Member of Congress

Barbara Mikulski
U.S. Senator

Constance Morella
Member of Congress

With the implementation of this law in 1994 requiring an analysis by sex and gender of outcomes, policies for the scientific design of clinical research began to provide the impetus for data-driven gender-specific medicine. While the NIH ORWH was not alone responsible for the change in scientific awareness of the importance of the role of sex and gender in biomedical research design, interpretation, and translational application, there is no doubt that the implementation of these new policies as requirements for funding provided a magnitude of incentive. To incorporate these concepts, intensive national strategic planning for priorities for the coming decades redefined the parameters of women's health research to encompass a better understanding of sex and gender differences in development, health, and disease.

Women's health priorities began to also call attention to factors such as the environment, lifestyle and behaviors, sex hormones, and social milieu, in addition to genetic inheritance, that could influence the susceptibility to genetic predispositions to diseases—what we now consider as epigenetic contributors.

Perhaps because of the early focus on clinical research, which is usually based on hypothesis first tested by basic research, further efforts were required to underscore the importance of determining sex differences in basic or preclinical research. The landmark 2001 Institute of Medicine report, *Exploring the Biological Contributions to Human Health: Does Sex Matter?*, documented known sex differences and strongly urged exploration of sex differences in basic cellular, molecular, and biochemical investigations, not just clinical research. It also emphasized advances in molecular biology that have revealed genetic and molecular bases for differences in health and human diseases, some of which result from the sexual genotype and therefore sex differences in health. It has been surprising that even today, there are investigators who are using cell lines in their research but had never considered if the cells are derived from males or females, or how the sex of the cells might affect their results! Fortunately, the ORWH announced a newly implemented policy in 2014 to require applicants to include in their research design their plans for the balance of male and female cells and animals in future grant applications for preclinical studies.

While principles of women's health were becoming part of the new norm of scientific design, another scientific area of discovery was taking place—that of defining the human genome. The goal of identifying and mapping the genes of the human genome was formally launched in 1990 and its completion was announced in 2003. While the "human genome" is considered a mosaic, the "genome" of each individual is considered unique. Although attention had been given to the genetic determination of health and disease and the role in health and disease of women, the revelations of the human genome, coupled with ever expanding studies of sex chromosomes, have brought about a marriage of modern concepts of biomedical research. That is, that not just behavioral or biological factors must be considered in defining women's health and gender-specific medical practices, but that the variations of the individual's genes, environment, and lifestyle must be taken into account for disease prevention and treatment. The Precision Medicine Initiative, implemented in 2015, is defined as an emerging approach for recommendations about prevention and treatment based on the individual's genes, environments, and lifestyles. The long-standing approach to sex- and gender-based studies and our understanding of the importance of factors including sex that can influence the penetrance of genetic predisposition for diseases seems to reflect the very principles which this new initiative is advancing.

However, with recent advances in our understanding of the sexual genotype, to the women's health community it has been disappointing that the definition of precision medicine does not specifically cite the consideration of the effects of biological sex on gene expression and therefore, noteworthy in precision medicine to determine recommendations for prevention or therapy.

Attention to sex and gender differences in etiology, presentation, diagnosis, treatment, or responses to therapy has become a norm of scientific design of research- and evidence-based medicine. Advances in our knowledge about contributions of each individual's genetic and chromosomal make up have further defined how research must define evidence that will be of great import for the clinical care of patients, regardless of their sex and/or gender. The ethical and social implications of genetic-based medicine are challenges that are being addressed. We have learned that it is necessary to examine variables of sex and gender across the spectrum of research, from basic molecular and cellular studies to clinical investigation, and ultimately to clinical application—providing personalized sex- and gender-appropriate health care to the individual patient through Gender-Specific Medicine.

CHAPTER 4

Gender Medicine in Austria

Alexandra Kautzky-Willer
Medical University of Vienna, Vienna, Austria

When I studied medicine at the Medical University of Vienna 1985 I did not learn anything about gender-specific medicine or even about the differences in women's or men's health other than aspects of gynecology and obstetrics for women and urology for men. When I finished medical school 1988 I decided to specialize in internal medicine because I wanted to study and cure humans from a holistic approach, covering a broad field of disease. I became especially interested in the field of endocrinology and metabolism, and started research in this particular topic at the end of my studies. I was fascinated by the complex impact of hormones on physiology and by how biology and psychology jointly shaped the human phenotype. Additionally, I was impressed by the gender-specific differences in endocrinologic physiology and disease.

When I became an Assistant Professor at the Medical University Clinic in Vienna, I soon developed a special interest in the field of diabetes. This is because it emerged as a very important topic: the prevalence rate is increasing and diabetes has an important impact on the quality of life and is associated with other serious illnesses. Diabetes has many faces: type 1 diabetes, which mostly occurs during childhood or adolescence, is an autoimmune disease. Type 2 diabetes, on the other hand, is clearly related to behavior and lifestyle factors with increasing incidence at older ages. I realized that diabetes impacts life and health of young women enormously, particularly with the need for special care during pregnancy and continuous monitoring after delivery. I became particularly interested in gestational diabetes, which is a category of its own in the spectrum of diabetic disease. This happened because I found that this disease was often undetected although it produced higher risk for both mother and child. Many clinicians were not

The International Society for Gender Medicine. DOI: http://dx.doi.org/10.1016/B978-0-12-811850-4.00004-1

even aware of it. In Austria there was no universal screening for gestational diabetes; hospitals had different criteria for diagnosis and no evidence-based treatment strategies. So I founded a national working group (Austrian Gestational Diabetes Working Group) and initiated studies to further explore the association of maternal glucose values and perinatal outcomes. We produced a better insight into the pathophysiology of impaired glucose metabolism, studying insulin sensitivity and secretion, endothelial function, inflammatory parameters, adipokines, and new biomarkers for different subgroups in the course of pregnancy and thereafter.

I became an elected member of the Diabetes and Pregnancy Study Group, part of the European Association for the Study of Diabetes (EASD). In 2004 I received the Joseph Hoet Research Award by the Diabetes Pregnancy Study Group. This award is given in recognition of excellence in research with important contribution to the advancement of knowledge in the field of diabetes and pregnancy. I was awarded for my pioneering work in advancing our understanding of the pathophysiology of gestational diabetes and the characterization of women with subtle cardiometabolic disturbances at risk of developing overt type 2 diabetes after delivery in longitudinal follow-up studies. At that time I realized that, because women with a history of gestational diabetes bear a very high risk of developing type 2 diabetes during 5–10 years after delivery, early risk stratification at the time of delivery could help to identify the women at highest risk. It became evident that such women could benefit from prevention programs and that both recurrence of gestational diabetes in a subsequent pregnancy as well as progression to overt type 2 diabetes could be avoided or delayed in about 60% of the cases. Moreover, women have an important impact on the health of their children and partners when changing their diet and increasing exercise, contributing to an increase in the well-being of the whole young family.

In 2007 the second congress of the International Society of Gender Medicine took place at the Medical University of Vienna, which was hosted by Professor Jeanette Strametz-Juranek, a cardiologist of the Medical University of Vienna, who founded the Austrian Society of Gender-Specific Medicine [1] as an interdisciplinary academic society in Vienna that same year. I was invited to give a lecture on the impact of gestational diabetes on women's health at the first congress of the Austrian Society of Gender-Specific Medicine in Baden, a small city near Vienna, 2008. This was the key moment to increase attention to gender medicine and encourage more investigation into this new discipline. Sex and gender differences play an important role in the pathophysiology, clinical picture, therapy, and complications of many disorders, particularly cardiometabolic diseases, which are increasing in incidence all over the world and which have a dramatic impact on the quality of life and life expectancy.

I was interested in the factors that produced type 2 diabetes and in particular in the distinct features of this disease in women with a history of gestational diabetes, compared to other women and men at risk for the disease. I wondered about the model of type 2 diabetes in men and started to follow a male cohort at risk to learn about potential sex-dimorphic pathophysiological processes.

In the recent past sex-specific aspects in diagnosis, therapy, and prevention of diseases were attracting more attention. The Council of Europe recommended including gender differences in health policy planning, delivery, and monitoring of health services [2].

Thus, this was the beginning of my engagement in the further development of the Austrian and thereafter of the International Society of Gender Medicine. After being part of the scientific advisory board, I was elected as board member of the Austrian Society in 2009 and in 2011 chair of this national society of gender-specific medicine.

Austria continues to focus more and more on gender-specific medicine. It is an interdisciplinary science, which encompasses important sex-dependent biological and gender-dependent differences. It incorporates the study of psychosocial differences in health behavior between men and women. The Medical University of Vienna expanded its role in this new field of research in the German-speaking area by establishing a professorship for Gender Medicine and in 2010, I became the first Chair of Gender Medicine in Austria and founded the Gender-Medicine Unit within the Department of Endocrinology and Metabolism. As newly established professor I set up and expanded a national and international scientific network, as well as an interdisciplinary Gender-Medicine facility at our University. Based on the multidisciplinary core area of endocrinology and metabolism, my aim was to encourage and establish interdisciplinary research projects and further develop teaching the evidence-based principles of gender medicine.

At our own university the research clusters and clinical special-focus programs were made aware of the new research approaches. Thus, new joint research projects were started and realized by means of cooperative collaborations. My vision was to link basic science and clinical research by means of interdisciplinary cooperation and to expand the science of sex and gender research in different subspecialties of medicine.

Indeed, endocrinology and especially diabetology turned out to be an ideal basis for sex- and gender-specific research, linking many diverse disciplines. Endocrinology and metabolism is a key field of sex- and gender-specific medicine [3]. Steroid hormones essentially contribute to differences between men and women, interacting with behavioral and environmental factors during lifetime, affecting health and illness. Almost all forms of autoimmune diseases, including endocrine forms like Hashimoto thyroiditis, Graves-Basedow disease, and Morbus Addison, show a much higher prevalence in females [4]. The underlying mechanisms of the heightened autoimmunity of women may be based on their more robust and reactive immune system with higher expression of inflammatory markers and the direct effects of sex hormones [5]. Testosterone exerts immunosuppressive reactions mediated by anti-inflammatory cytokine production and inhibition of gene expression implicated in immune activation [6]. In contrast, estrogens are associated with inflammatory stimuli and immune cell activation and differentiation.

Type 1 diabetes represents an exception among autoimmune disorders with an overall balanced sex ratio but with striking sex differences [7,8]. Higher prevalence was reported in young girls until puberty but in adolescence and throughout reproductive age more males than females were affected [9]. In addition, girls showed a higher rate of severe diabetic ketoacidosis [10]. In contrast to the situation for women, the risk for end stage renal disease and proliferative retinopathy were doubled in men if diabetes onset occurred after 15 years of age [11,12]. Although the reason for these age- and sex-dependent differences in prevalence rates of type 1 diabetes and diabetic complications are unclear, genetic factors, sex hormones, and environment may interact in a sex-specific way, resulting in different immune response and organ vulnerability between men and women.

Gender dimorphism is also relevant in all kinds of thyroid diseases with a clear preponderance in women, as well as in neuroendocrinology and bone metabolism [7,8]. However, the greatest body of evidence for important clinical implications comes from studies in the field of obesity and diabetes: diabesity [3,13]. Genetic background, lifestyle, and environment contribute to the pandemic increase of diabesity, presenting a challenge to health care systems.

In 2014, Margarethe Hochleitner, who was Director of the Women's Health Centre at the Innsbruck University Hospital in Tyrol, became Professor for Gender Medicine at the Medical University of Innsbruck. Her research focus comprised cardiology, preventive medicine, gender studies, women's health, and the health of migrants.

At both universities, gender medicine is now incorporated into the mandatory curriculum aiming to teach the students skills in the field of gender medicine. Moreover, regular lecture series on topics relating to gender medicine are offered, which students can complete as part of their elective subjects. The increasing number of diploma and doctoral theses based on sex and gender research topics shows the increasing interest and demand for gender medicine at both medical universities. It is also a good example of combining teaching and research in a synergistic way. Both universities are now starting a curriculum for a diploma of gender medicine in collaboration with the Austrian Chamber of Medicine for Physicians in order to increase clinicians' knowledge of the most prominent sex and gender differences of common diseases and thus to improve patient care [14].

In addition, at the Medical University of Vienna, throughout the course of (life-long) university education, we aimed at conveying major competencies in the field of gender medicine to our students by offering the first comprehensive postgraduate course in gender-specific medicine in Europe [15]. It offers a comprehensive survey of biopsychosocial approaches to women's and men's health, as well as gender-sensitive conduct and clinical skills. Practical insights are combined with the findings of cutting-edge science and research and gendered innovations. The curriculum is based on international standards and guidelines and takes an evidence-based approach. It provides students with know-how, abilities, and professional conduct skills required to improve the long-term health of women and men. Course content covers the professional principles and knowledge required in the fields of health care, clinical practice, research, education, training, and health care policy. This is essential in order to more fully meet the individual, gender-specific health care needs of women and men and therefore make lasting improvements in quality. Demand for the gender-medicine competence has grown—in basic research, specialized medical disciplines, public health, and health care policy.

The estimation of 9% of severe obesity in women and 6% in men in 2025 [16] shows the urgent need for effective prevention. Otherwise the increase of obesity-related complications, like type 2 diabetes, depression, and cardiovascular disease as well as cancers, will lead to lower Health-Related Quality of Life (HRQL) and increased mortality rates in both sexes, however especially in women.

In particular, diabetes is a much greater risk factor for cardiovascular disease in women compared to men [3]. The sex-dimorphic effects already start in utero, as shown by epidemiological data. Therefore diabesity prevention must start in utero, targeting women planning pregnancy as well as early pregnancy.

The importance of diabetes, in regard to sex-specific aspects and public health, is evidenced by the fact that the main topic of the world diabetes day 2017, of the International Diabetes Federation, is dedicated to women and diabetes [17]. Up to 70% of cases of type 2 diabetes could be prevented through the adoption of a healthy lifestyle, 70% of premature deaths among adults are largely due to behavior initiated during puberty. Diabetes is the ninth leading cause of death in women, two out of every five women with diabetes are of reproductive age. Women with diabetes are almost 10 times more likely to have coronary heart disease than women without the condition and additionally they have an increased risk of early miscarriage or having a baby with malformations. One in seven births is affected by gestational diabetes.

Therefore our research still has one main focus, which is on diabetic women of reproductive age. Our work contributed to the inclusion of the oral glucose tolerance test between the 24th and 28th gestational week in the mother- child booklet, to enable universal screening in all pregnant women in Austria [18,19]. In addition, we participated in EU projects seeking possibilities to prevent gestational diabetes and its sequelae in obese women [20]. We also collaborate in other EU projects, aiming to increase gender awareness in cardiovascular diseases [21].

At the moment our research activities cover a broad spectrum of complex diseases affecting men and women in different ways. We conduct cohort studies in men and women at risk for diabetes, studying early sex-specific biomarkers in order to develop sex-sensitive risk scores. We want to learn more about sexual dimorphism in the pathophysiology of metabolic diseases and their complications throughout life cycle, trying to establish individual gender-sensitive prevention strategies and therapeutic concepts based on new findings. We try to link basic science and clinical studies in sex and gender research to translational and precision medicine. We are also interested in achieving a better insight into sex-dimorphic epigenetic effects of diet, lifestyle, and environment. Overall, we aim to get a broader gender-sensitive knowledge providing the basis for specific evidence-based interventions, allowing prevention, treatment, rehabilitation, and health care delivery strategies to match the different needs for women and men.

Fields of translational research termed "-omics" (genomics, proteomics, and metabolomics) study the contribution of genes, proteins, and metabolic pathways to human physiology and the variations of these pathways that can lead to disease susceptibility. Hoping that these fields will enable new and better approaches of diagnosis, drug development, and individualized therapy, we are proposing a new "-omics" (*genderomics*) related to sex differences. Its aims are the systematic scientifically based gender analyses, the promotion of large prospective research studies on the impact of gender on primary outcomes, and the implementation of evidence-based sex-/gender-specific clinical recommendations. We look forward to further research on sexually dimorphic pathophysiological mechanisms (essential to further improve gender awareness) and to improve the competency in the health care system in the principles of gender medicine.

It is necessary that all scientific associations and research communities are aware of the need for sex-specific analysis and sex-sensitive recommendations. To this end we included a chapter of sex-specific considerations and recommendations in the guidelines of the Austrian Diabetes associations [22]. Also the American Heart and Stroke Association published specific recommendations for diabetic women, targeting differences in risk factors, diagnostic procedures, and drug therapies [23,24]. Efforts are needed to incorporate symposia and workshops on gender medicine in the annual meetings of the specific medical societies, to better integrate sex and gender research, and clinical practice in the various disciplines.

Although much progress was achieved in the last decade there is still a long way to go and there are still many obstacles and black boxes. However, increased awareness of health professionals, patients, general public, and health policy regarding sex and gender in development and management of most diseases is necessary and worthwhile. Both basic research and clinical studies give evidence of many important sex and gender differences, but there are also many controversial issues and open questions. Many results are derived from observational and association studies, and do not investigate basic mechanisms underlying their data. Moreover, we need large randomized controlled trials proving sex-specific effects by adequately designed interventions and long-term follow-ups, including the sociocultural dimension of gender. Additionally, appropriate animal models and translational research to study sex differences are needed to get more insights into the pathophysiology and complex interplay of genes, hormones, lifestyle, and environment.

I am confident that future considerations of female cells and research animal models will increase. Unfortunately, however even recent reports in cardiovascular research—an area of a long tradition in gender medicine—showed that gender bias persists in preclinical research and that the number of women included in trials is insufficient and did not improve significantly in recent years. Nevertheless, an inclusion of an adequate proportion of women in clinical studies is essential for the promotion of sex-specific analysis. Future biomedical research based on gender-based applications could contribute to a better reproducibility of research.

In my vision, the future of gender medicine in Europe will be the end of the "one-size-fits-all medicine." Sex- and gender-based medicine will be integrated in clinical care and help to improve individual prevention and care of diseases. Gender-based medicine will not at all become irrelevant with the implementation of personalized medicine but will be synergistic. On the contrary, sex- and gender-specific medicine will be an important part, an enlargement of personalized or precision medicine. Analysis of large databases prove that sex and gender persist as independent and remarkable risk factors because some genetic variants mediate sex-dimorphic risks, independent of other factors including age and comorbidities. Additionally, based on differences in biology, gender medicine brings social, cultural, and behavioral elements to health issues.

In 2015, I received the national award of gender studies by the Federal Ministry of Science, Research and Economy, called the Possanner von Ehrenthal Award, who was the first female doctor in Austria in 1900. In 2017 I was elected as "scientist of the year" in Austria by the Austrian journalists of education and science [25]. I take these honors as signs for the increasing appreciation of gender medicine in my country. I hope that we will succeed to continue to enlarge our scientific networks and research possibilities and to establish guidelines for the implementation of gender medicine into clinical care.

REFERENCES

[1] Österreichische Gesellschaft für geschlechtsspezifische Medizin. Available from: http://www.gendermedizin.at/ [accessed 05.02.15].

[2] Committee of Ministers. 6.2 European Health Committee (CDSP) Draft Recommendation CM/Rec(2007)... of the Committee of Ministers to member states on the inclusion of gender differences in health policy. Council of Europe; 2007 [updated 26.10.07]; Available from: <https://wcd.coe.int/wcd/ViewDoc.jsp?id=1202389>; [accessed 18.05.11].

[3] Kautzky-Willer A, Harreiter J, Pacini G. Sex and gender differences in risk, pathophysiology and complications of type 2 diabetes mellitus. Endocr Rev 2016;37(3):278–316.

[4] Ober C, Loisel DA, Gilad Y. Sex-specific genetic architecture of human disease. Nat Rev Genet 2008;9(12):911–22.

[5] Sakiani S, Olsen NJ, Kovacs WJ. Gonadal steroids and humoral immunity. Nat Rev Endocrinol 2013;9(1):56–62.

[6] Furman D, Hejblum BP, Simon N, Jojic V, Dekker CL, Thiébaut R, et al. Systems analysis of sex differences reveals an immunosuppressive role for testosterone in the response to influenza vaccination. Proc Natl Acad Sci USA 2014;111(2):869–74.

[7] Kautzky-Willer A, Lemmens-Gruber R. Obesity and diabetes. In: Regitz-Zagrosek V, editor. Sex and gender differences in pharmacology. Berlin: Springer; 2012. p. 307–40.

[8] Kautzky-Willer A. Sex and gender differences in endocrinology. In: Oertelt-Prigione S, Regitz-Zagrosek V, editors. Sex and gender aspects in clinical medicine. London: Springer; 2012. p. 125–49.

[9] Rami B, Waldhör T, Schober E. Incidence of type I diabetes mellitus in children and young adults in the province of Upper Austria, 1994–1996. Diabetologia 2001;44(Suppl. 3):B45–7.

[10] Schober E, Rami B, Waldhoer T. Diabetic ketoacidosis at diagnosis in Austrian children in 1989-2008: a population-based analysis. Diabetologia 2010;53(6):1057–61.

[11] Möllsten A, Svensson M, Waernbaum I, Berhan Y, Schön S, Nyström L, et al. Cumulative risk, age at onset, and sex-specific differences for developing end-stage renal disease in young patients with type 1 diabetes: a nationwide population-based cohort study. Diabetes 2010;59(7):1803–8.

[12] Harjutsalo V, Maric C, Forsblom C, Thorn L, Waden J, Groop PH. Sex-related differences in the long-term risk of microvascular complications by age at onset of type 1 diabetes. Diabetologia 2011;54(8):1992–9.

[13] Kautzky-Willer A, Handisurya A. Metabolic diseases and associated complications: sex and gender matter!. Eur J Clin Invest 2009;39(8):631–48.

[14] Österreichische Ärztekamer. ÖÄK-Diplom Gender Medicine. 2014 [updated 17.09.14]; Available from: https://www.arztakademie.at/oeaeknbspdiplome-zertifikate-cpds/oeaek-spezialdiplome/gender-medicine/; [accessed 20.05.17].

[15] Medical University of Vienna. Postgraduate university course Gender Medicine. Vienna: Medical University of Vienna; 2010 [updated 07.03.17]; Available from: <https://www.meduniwien.ac.at/hp/1/ulg-gendermedicine/>; [accessed 20.05.17].

[16] NCD Risk Factor Collaboration (NCD-RisC). Trends in adult body-mass index in 200 countries from 1975 to 2014: a pooled analysis of 1698 population-based measurement studies with 19.2 million participants. Lancet 2016;387(10026):1377–96.

[17] Piemonte L. IDF announce WDD 2017 theme on International Women's Day. International Diabetes Federation; 2017 [updated 08.03.17]; Available from: <http://www.idf.org/news/idf-announce-wdd2017-theme-international-womens-day?language=en>; [accessed 19.05.17].

[18] Kautzky-Willer A, Harreiter J, Bancher-Todesca D, Berger A, Repa A, Lechleitner M, et al. Gestationsdiabetes (GDM). Wien Klin Wochenschr 2016;128(Suppl. 2):103–12.

[19] Kautzky-Willer A, Bancher-Todesca D, Weitgasser R, Prikoszovich T, Steiner H, Shnawa N, et al. The impact of risk factors and more stringent diagnostic criteria of gestational diabetes on outcomes in central European women. J Clin Endocrinol Metab 2008;93(5): 1689–95.

[20] Simmons D, Devlieger R, van Assche A, Jans G, Galjaard S, Corcoy R, et al. Effect of physical activity and/or healthy eating on GDM risk: the DALI Lifestyle Study. J Clin Endocrinol Metab 2016;jc20163455.

[21] The EUGenMed, Cardiovascular Clinical Study Group, Regitz-Zagrosek V, Oertelt-Prigione S, Prescott E, Franconi F, et al. Gender in cardiovascular diseases: impact on clinical manifestations, management, and outcomes. Eur Heart J 2016;37(1):24–34.

[22] Kautzky-Willer A, Abrahamian H, Weitgasser R, Fasching P, Hoppichler F, Lechleitner M. Geschlechtsspezifische Aspekte für die klinische Praxis bei Prädiabetes und Diabetes mellitus. Wien Klin Wochenschr 2016;128(Suppl. 2):151–8.

[23] Bushnell C, McCullough L. Stroke prevention in women: synopsis of the 2014 American Heart Association/American Stroke Association guideline. Ann Intern Med 2014;160(12): 853–7.

[24] Regensteiner JG, Golden S, Huebschmann AG, Barrett-Connor E, Chang AY, Chyun D, et al. Sex differences in the cardiovascular consequences of diabetes mellitus: a scientific statement from the American Heart Association. Circulation 2015;132(25):2424–47.

[25] Lehmann O. Gendermedizinerin Kautzky-Willer ist "Wissenschafterin des Jahres". Klub der Bildungs- und WissenschaftsjournalistInnen; 2017 [updated 07.01.17]; Available from: http://www.wissenschaftsjournalisten.at/2017/01/09/gendermedizinerin-kautzky-willer-ist-wissenschafterin-des-jahres/; [accessed 26.05.17].

CHAPTER 5

Gender Medicine in Cardiovascular Diseases: Past, Present, and Future

Gender Medicine—What Else?

Jeanette Strametz-Juranek[1], Martin Skoumal[1], Sandra Steinböck[2] and Angelika Hoffer-Pober[2]

[1]Pensionsversicherungsanstalt, Graz, Austria [2]Medical University of Vienna, Vienna, Austria

INTRODUCTION

Cardiovascular diseases (CVD) are the leading cause of death worldwide [1–3]. In 2015 47% of women and 37% of men were dying because of CVD in Austria [4]. In general women are about 10 years older compared to men having their first event and tend to have a higher 30-day mortality after myocardial infarction (MI) [4]. This increased mortality appears to be limited to ST elevation refers to "finding on an electrocardiograph wherein the trace in the ST segment is abnormally high above the baseline". ST elevated MI and can to some extent be explained by the different clinical symptoms at the time of presentation [5,6]. Women have a poorer prognosis after coronary bypass graft surgery [7], reflecting technical difficulties resulting from different risk factor profiles and smaller body size. In general, death rates after MI have declined in the last 30 years due to improved technical as well medical treatments [8–14]. However, death rates have not improved as much in women compared to men. One reason for this might be coronary microvascular disease (MVD), a new concept affecting the smallest heart vessels, limiting or preventing oxygen-rich blood flow to the myocardium [15]. The Women's Ischemia Syndrome Evaluation Study (WISE) study has shown that MVD affects about 3 million women with coronary heart disease in the United States [10]. MVD seems to be a challenge for doctors because standard tests for CVD are only useful in finding blockages in large coronary arteries. However, women with MVD have "clean" arteries in angiography and can only be diagnosed with intravascular ultrasound.

The International Society for Gender Medicine. DOI: http://dx.doi.org/10.1016/B978-0-12-811850-4.00005-3

MVD is a very wonderful example of the different needs of women and men with heart disease. It illustrates the importance of an interdisciplinary collaboration and the importance of having a gender specific perspective in cardiology.

Although a majority of differences in CVD can be explained to be sex-based (i.e., steroid hormones and anatomy), gender-related factors such as culture, language, religion, education, income, life style, emotions, health behavior, awareness of disease, and access to health care are also important factors influencing outcome and prognosis [16,17]. Men and women tend to deal differently with life-changing circumstances and heart diseases are no exception. However, a majority of gender issues are still barely addressed in medicine [18,19].

Over the past decades scientists, health-care providers, and policy-makers have made substantial effort to improve the understanding of gender-specific differences in CVD. However, there is still a long way to go to close the gender gap in medicine.

In the following chapter we will describe the beginning of gender-specific medicine (GSM) in Austria. We will trace the development of different educational activities, research, and the activities of scientific and lay communities. We will focus on the inspiring future perspective of implementing GSM in the Master Plan Rehabilitation of the Pensionsversicherung in Austria. This program is structured to ensure that all patients, both during their stay as an inpatient or during ambulatory rehabilitation, will receive their treatment not only according to the International Classification of Functional Disability but also based on a gender-specific perspective [20].

INSTITUTIONAL BACKGROUND—GENDER MAINSTREAMING AND DIVERSITY AT THE MEDICAL UNIVERSITY OF VIENNA

Gender Mainstreaming at the Medical University was implemented in 2004. The competence portfolio of the Gender Mainstreaming Office supports equal opportunities and measures for the advancement of women as well as the integration of gender as a cross-cutting issue in research and education. This focus on both—equal opportunities and gender studies/gender medicine—follows the Austrian University Law. The Gender Mainstreaming Office renamed in 2017 "Gender Mainstreaming and Diversity" supports and advises the Rectorate and all department and office heads in issues relating to equal opportunities and the advancement of women. It particularly deals with the elimination of structural barriers for women researchers at the Medical University of Vienna and implements measures to increase the proportion of women at all levels, especially in executive positions.

Specific programs and measures for the advancement of women are devised to encourage them to pursue a scientific career and support them in this career path. The Office thus designs, coordinates, and supports all organizational units in developing and implementing specific measures and programs for women which seek to establish a balanced number of women and men at all levels. It also strives to eliminate the existing underrepresentation of and discrimination against women and to strengthen the professional standing of women in research.

GENDER-SENSITIVE RESEARCH AND TEACHING

The Department of Gender Mainstreaming and Diversity supports the Rectorate in implementing the tasks specified in the interdisciplinary research field Women's Health and Gender-Based Medicine, which forms part of the curriculum of the degree program "Human Medicine." The work of the department supports pooling and cross-linking activity in the field of women and gender research at the Medical University of Vienna. It promotes conferences, symposia, and congresses, national and international networking, and research on the topic of women in research. The goal is to continuously reflect on the work carried out by the Office and to create a foundation for drawing up specific recommendations for gender equity. This includes, for instance, researching the concrete situation of female employees as well as a retrospective data analysis of specific topics like gender medicine research. Another task of the Department is to consolidate, coordinate, and connect existing research activities in the field of women's and gender studies at the Medical University of Vienna.

Gender Lectures

In the course of (life-long) university education, the Medical University of Vienna aims to convey competencies in the field of gender medicine to its students. The Department of Gender Mainstreaming and Diversity organizes two lecture series with 3-hour lectures on topics in gender medicine per semester and a lecture series with 3-hour lectures on topics in the category of "diversity and medicine" every second semester. These courses can be completed as free elective subjects. The Medical University of Vienna also seeks to incorporate gender medicine into the mandatory curriculum. For this purpose, the Department of Gender Mainstreaming and Diversity regularly meets with lecturers at the Medical University of Vienna in the working group "Gender in the Curriculum" (founded in 2006 and now "Gender and Diversity in the Curriculum") to make them aware of the topic and support and advise them. Participants (main stakeholders teaching in different fields of expertise) receive professional support in designing and implementing strategies for integrating gender-specific aspects in mandatory classes. By May 2010, six half-day workshops had taken place in which a joint conclusion paper was drawn up. Since then, workshops have been organized twice a year to monitor the implementation of the conclusion paper and develop new measures.

Postgraduate Program Gender Medicine

Scientific research of biological and psychosocial differences between women and men in the development, perception, and treatment of illnesses and the definition of gender-specific forms of treatment have developed into an interdisciplinary field of its own during the last years: gender medicine. As the first Austrian university, the Medical University of Vienna has offered a postgraduate program on gender medicine since the winter semester of 2010. In 2010, The Medical University of Vienna established a professorship in gender medicine.

GENDER MEDICINE IN CARDIOVASCULAR DISEASES: THE PAST

Education and Research

In 2004 I was invited by my college and friend Brigitte Litschauer, the former head of the Department of Gender Mainstreaming, a physiologist and expert on stress disorders, who was mostly responsible for implementing GSM in the medical curriculum of the Medical University of Vienna, Austria, to assemble gender-specific lectures for medical students at the university. They were called gendeRingvorlesungen. In the beginning it was really very hard to find scientists and doctors for these seminars, but we succeeded and we started in 2005 with eight students. One seminar focused on a specific topic, the other on an interdisciplinary medical field. Over the years more and more colleagues collaborated with us and we had the opportunity to build a national gender network all over Austria with an increasing number of gender experts and students who were interested in this topic. As a result of the seminar, students began to ask for the opportunity to compose their doctoral thesis on gender-specific issues in cardiology and since 2014 I had the opportunity to supervise 24 medical students. Out of these, 14 were supported by a grant from the university and 2 students received the distinction of the Lore Antoine Prize from the Austrian Medical Women's Association. Further, two young medical doctors decided to choose a gender topic for their master's thesis. One of them worked on CVD and diabetes and the other on cardiovascular awareness and prevention. Both received specific grants for their work on GSM by the Austrian Research Promotion Agency of the Austrian Ministry for Transport, Innovation and Technology. In total we were supported with about 450,000 Euros. Moreover, the Niederösterreichische Gebietskrankenkasse, a public insurance company in Austria, invited us to collaborate and to investigate cardiovascular awareness, preventive action, and barriers to cardiovascular health in Lower Austria. The company supported this project with 150,000 Euros.

Although in 2014 I moved from the Medical University of Vienna to become the head of the Sonderkrankenanstalt Rehabilitationszentrum Bad Tatzmannsdorf, which is a large rehabilitation center for CVD in Burgenland, gender-specific research is still going on. Recently I had the opportunity to supervise a master's thesis of one of the doctors at the Medical University of Graz working in our center focusing on the gender-specific aspects of nonalcoholic fatty liver and, CVD. She successfully defended her thesis in April 2017. Another project concerns the role of malnutrition in cardiovascular rehabilitation and its influence on participation in rehabilitation programs from a gender-sensitive perspective. This project is part of a doctoral thesis and will start in June 2017.

Foundation of the Austrian Society of Gender Specific Medicine (ÖGGSM) and the International Society for Gender Medicine

In 2007 the Austrian Society of Gender Specific Medicine was founded by Adelheid Gabriel, Gabriele Fischer, Anton Luger, Michael Eisenmenger, Thomas Juranek, and myself. The society is organized as an interdisciplinary, academic society and is a platform for physicians, scientists, and professionals who are interested in GSM and who share the common goal of promoting this science. In 2008 the ÖGGSM started to organize scientific sessions and symposiums. A highlight was in April 2010, when the ÖGGSM had the privilege of organizing the international symposium "Global health and gender" in cooperation with the nongovernmental organizations of the United Nations (UN), introducing GSM at the level of the UN. The ÖGGSM could achieve collaboration with national societies, that is, the Austria Society of Cardiology, the Austrian Society of Urology, Austrian Medical Women's Association, and the Austrian Society for Man and Health. Of course the society is also part of the International Society for Gender Medicine (IGM).

The IGM is an umbrella society for organization for national and professional societies, dedicated to the study of gender- and sex-specific differences. The society was founded in 2006 by Marianne Legato (New York), Maria Grazia Modena (Modena), Vera Regitz-Zagrosek (Berlin), Karin Schenck-Gustafsson (Stockholm), and myself. Within a short time, IGM has gained a worldwide reputation, has sponsored seven international congresses, has promoted research and teaching, and has indeed "put gender on the agenda."

GENDER MEDICINE IN CARDIOVASCULAR DISEASES: THE PRESENT

In August 2014 I took over the medical management of the Sonderkrankenanstalt (SKA) Rehabilitationszentrum Bad Tatzmannsdorf of the Pensionsversicherung in Austria. Our interdisciplinary team consists of medical doctors, physiotherapists, dieticians, psychologists, and nurses and takes care of 172 patients with CVD. We also can offer the opportunity of cardiovascular rehabilitation to patients with a prolonged course of disease or increased medical or nursing expenses on our ward.

Our rehabilitation program is an inpatient medically supervised program focusing on exercise counseling and training, education for heart-healthy living, counseling to reduce stress and anxiety, counseling for the retention of professional capacity and return to work. Patients enjoy case and care management based on the International Classification of Functional Disability Criteria (ICF) [20]. Medicare and most public as well as private insurance companies cover a standard inpatient cardiac rehabilitation program that includes 120–135 therapy minutes/day over 3–4 weeks. After this period of time employed patients have the opportunity of continuing the rehabilitation program in an ambulatory rehabilitation center. Special interdisciplinary training courses in our center are performed for patients with diabetes and heart failure. A special smoking cessation program is offered to all patients willing to stop smoking during their stay in our clinic. Our clinic is also a member of the International Network of Health Promoting Hospitals and Health Services, which was initiated by the World Health Organisation, and in 2016 has been named as the first silver-certified smokeless health department in Burgenland.

In 2015 we began to implement gender-specific topics into our therapies and courses for the patients and we started in autumn 2015 with a seminar for all new patients about the importance of cardiovascular rehabilitation and the differences between women and men in CVD and its risk factors. This occurs weekly. We also implemented GSM into our courses and therapies for dieticians, training therapists, and psychologists. Our team wrote a booklet "Wegweiser durch die geschlechtsspezifische Rehabilitation," giving all patients information about treatment and therapy performed in our center. We stressed the importance of individual responsibility and the different needs of women and men with CVD. We are currently revising our training materials for patients on various gender-specific topics. To ensure that our team has current knowledge in GSM we conduct regular training with gender medicine experts. It is fantastic to see how GSM expertise is growing and how much it is inspiring our work with the patients.

GENDER MEDICINE IN CARDIOVASCULAR DISEASES: THE FUTURE

The Pensionsversicherungsanstalt is one of the largest providers of rehabilitation in Austria and operates 15 inpatient rehabilitation clinic centers throughout the country for CVD, metabolic, musculoskeletal, oncologic-, neurologic-, and pulmonary diseases. There are two outpatient rehabilitation centers in Vienna and Graz. Table 5.1 shows the number of patients treated in all of the 15 rehabilitation clinics of the Pensionsversicherung in 2016.

In 2016 Dr. Martin Skoumal became the head physician and together with the executive board of the Pensionsversicherung the Master Plan Rehabilitation was designed in November 2016 to renew

Table 5.1 Underlying Diseases and the Referral to Rehabilitation 2016

Underlying diseases *n* =	Male	Female	Total
Gastrointestinal	272	266	538
Cardiovascular	6.677	2.796	9.473
Lung	1.352	1.011	2.363
Morbus Bechterew	261	126	387
Neurologic	1.013	755	1.768
Metabolic	2.409	1.598	4.007
Obesity	145	72	217
Amputation	53	28	81
Movement and support apparatus	5.513	6.911	12.424
Total	17.804	13.858	31.662

the concept of rehabilitation in Austria. The Master Plan has four cornerstones:

1. The revised *Medical Performance Profiles* (*MLP*) provides therapies and treatment in three performance categories to assure that all patients receive exercise and strength training protocols according to their individual capacity. We ensure regeneration time between therapies. The existing medical performance profiles were focused on the needs of men—this means that there is no consideration of the impact of sex, age, concomitant diseases, exercise capacity, or mobility.
2. Rehabilitation based on *International Criteria of Functional Disability* (*ICF*)—the ICF classification complements WHO's International Classification of Diseases—10th Revision (ICD), which contains information on diagnosis and health condition, but not on functional status [20]. The ICF is structured around the following broad components:
   - body functions and structure
   - activities related to tasks and actions by the individual person and participation in her/his life situation
   - additional information on severity and environmental factors

   Functioning and disability are viewed as a complex interaction between the health condition of the individual and the contextual factors of the environment as well as personal factors. The classification maintains that these factors are interactive and dynamic rather than linear or static. It allows for an assessment of the degree of disability, although it is not a measurement instrument.

3. *Preservation of employability and return to work*: one of the main tasks of rehabilitation is to enable the patients to live their lives independently, and to enable them to work or to start or complete their education. Disabled retirement or disability-required care should be avoided as long as possible or at least postponed. So the return to work process will start for the patients early at the rehabilitation clinic with special courses (training and exercise, psychology, dietetics, and social law) for employed people. This gives patients as soon as possible information about the possibility of their returning to a work place and whether or not adjustments are needed. In case of adjustments or occupational retraining the Pensionsversicherung will also help to initiate this process.
4. *Case and care management*: Each rehabilitation clinic of the Pensionsversicherung should have a case and care manager, who works closely with the patients who need chronic care. This gives his/her family and the physicians a plan to structure home care or to utilize nursing homes and assisted life facilities. These managers begin their work soon after the patient is admitted to the rehabilitation clinic.

Together with Martin Skoumal, we conceived the idea of integrating GSM into the Master Plan Rehabilitation (MLP, ICF, preservation of employability and return to work and case and care management) to ensure that in every part of this program gender-specific aspects are considered in an adequate way, reflecting the different needs and requirements of women and men in rehabilitation. Together with Sandra Steinböck and Angelika Hoffer-Pober of the Department of Gender Mainstreaming and Diversity from the Medical University of Vienna, Austria, we are now working on different intervention levels and projections in the four cornerstones of the Master Plan. We are assembling the gender expertise of the interdisciplinary members of the different rehabilitation teams to create a gender-based medical history questionnaire and a guideline for a gender suitable conversation, creating gender-specific knowledge in diagnosis and treatment. Moreover we also want to investigate differences between women and men concerning referral to rehabilitation, the length of stay, and the degree to which they achieve their rehabilitation aims.

Thus, the Master Plan Rehabilitation of the Pensionsversicherung will be the first rehabilitation program on a national as well as an international level based on GSM. It is inspiring and an honor to be a part of this team working on it.

REFERENCES

[1] World Health Organization. Global status report on noncommunicable diseases 2014. Available from: http://www.who.int/mediacentre/factsheets/fs317/en/ [accessed 19.05.17].

[2] Benjamin EJ, Blaha MJ, Chivue SE, et al. Heart disease and stroke statistics—2017 update. A report from the American Heart Association. Circ 2017 Jan. Available from: http://dx.doi.org/10.1161/CIR.0000000000000485.

[3] Statistik Austria. Todesursachenstatistik. 2016. http://www.statistik.at (in German).

[4] Coulter SA. Epidemiology of cardiovascular disease in women. Risk, advances, and alarms. Tex Heart Inst J 2011;38(2):145–7.

[5] Hochman JS, Tamis JE, Thompson TD. Global use of strategies to open occluded coronary arteries in acute coronary syndromes II b investigators. Sex, clinical presentation and outcomes in patients with acute coronary syndromes. N Engl Med 1999;341:226–32.

[6] Berger JS, Elliott L, Fallup D. Sex differences in mortality following acute coronary syndromes. JAMA 2009;302:874–82.

[7] Aldea GS, Gaudiani JM, Shapira OM. Improved clinical outcomes in patients undergoing coronary bypass grafting with coronary endartherectomy. Ann Thorac Surg 1999;67:1097–103.

[8] Bello N, Mosca L. Epidemiology of coronary heart disease in women. Prog Cardiovasc Dis 2004;46(4):287–95. Review. PubMed PMID: 14961452.

[9] EUGenMed Cardiovascular Clinical Study Group, Regitz-Zagrosek V, Oertelt-Prigione S, Prescott E, Franconi F, Gerdts E, et al. Gender in cardiovascular diseases: impact on clinical manifestations, management, and outcomes. Eur Heart J 2016;37(1):24–34. Available from: http://dx.doi.org/10.1093/eurheartj/ehv598. Review. PubMed PMID: 26530104.

[10] Bairey Merz CN, Johnson BD, Sharaf BL, et al. Hypoestrogenemia of hypothalamic origin and coronary artery disease in premenopausal women: a report from the NHLBI-sponsored. WISE study. J Am Coll Cardiol 2003;41:413–19.

[11] Herz-Kreislauf-Erkrankungen in Österreich: Angina pectoris, Myokardinfarkt, ischämischer Schlaganfall, periphere arterielle Verschlusskrankheit. Epidemiologie und Prävention. Bundesministerium für Gesundheit. 2014 (in German).

[12] Maas AHEM, Appelman YEA. Gender differences in coronary heart disease. Neth Heart J 2010;18:598–603.

[13] Hochman JS, Tamis JE, Thompson TD, et al. Sex, clinical presentation and outcome in patients with acute coronary syndromes. N Engl J Med 1999;341:226–32.

[14] Mosca L, Barrett-Corner E. Sex/gender differences in cardiovascular disease prevention: what a difference a decade makes. Circ 2011;124:2145–54.

[15] Shaw LJ, Bugiardini R, Merz NB. Women and ischemic heart disease: evolving knowledge. J Am Coll Cardiol 2009;54(17):1561–75.

[16] Mosca L, Mochari-Greenberger H, Dolor RJ. Twelve-year follow-up of American women's awareness of cardiovascular disease risk and barriers to health care. Circ Cardiovasc Qual Outcomes 2010;3(2):120–7.

[17] Haidinger T, Zweimüller M, Stütz L, et al. Effect of gender on awareness of cardiovascular risk factors, preventive action taken and barriers to cardiovascular health in a group of Austrian subjects. Gender Med 2012;9(2):94–102.

[18] Cameron E, Bernards J. Gender and disadvantage in health: men's health for a change. Soc Health Illness 1998;20(5):637–93.

[19] Baker P, Dworkin SL, Tong S, et al. The men's health gap: men must be included in the global health equity agenda. Bull World Health Organ 2014;92:618–20. Available from: http://dx.doi.org/10.2471/BLT.13.132795.

[20] World Health Organisation. International Classification of Functioning, Disability and Health (ICF). Available from: http://www.who.int/classifications/icf/icf_more/en/ [accessed 19.05.17].

CHAPTER 6

Sex and Gender in Health: The World Writes on the Body

Gillian Einstein
Department of Psychology, University of Toronto, Toronto, ON, Canada

INTRODUCTION

As were many of my generation, I was slow to wake up to the fact that what we knew about biological science was only about males or that there might be any differences between males and females. No one ever mentioned it in my courses—even in graduate school. I genuinely believed that the use of the pronoun, "he," was a reference to both females and males/women and men. I focused more on the difficulty of being a woman in science, reading biographies of women scientists trying to get a grip on how one would be that.

One day, however, during an undergraduate class, I got the picture that there was a vast ignorance about women's health and their physiology, exclusive of their reproductive biology. In a course on public health, one, particular case study caught my attention. The Professor kept referring to the participants as, "he," which was the general pronoun then. but I wondered: Could these findings be applied to women as well as men? I was genuinely curious because the repercussions for women could have been different than those for men. I approached the Professor after class and asked if the findings were about both females and males. He said, "yes," but was so visibly taken aback by the question that I realized he did not know. That was 1975.

In 1976, The Boston Women's Health Collective published, "Our Bodies, Ourselves." [1] It outlined women's experiences in biomedicine, especially around reproductive health, a condition that back then was viewed as a disease but is, in fact, normal for women. The handbook was about treating pregnancy as a healthy part of a woman's life and about women's taking ownership of their own reproductive health and body.

The International Society for Gender Medicine. DOI: http://dx.doi.org/10.1016/B978-0-12-811850-4.00006-5

This was the first time I had heard about women's bodies and health-care experiences being different from those of men's. The first edition focused on reproductive health since little was known about women's other body systems as distinct from men's. After reading "Our Bodies, Ourselves," I knew the professor actually did not know what the case was for females—or if the cited study even included women. The quandary made a lasting impression on me.

This experience awakened me to the fact that we needed to learn if the bodies and lives of females/women were different from males/men and if the experiences of those bodies were different, how they had an impact on biology.

Encapsulated in this story are some of the issues that are relevant for the field of gender-specific medicine (GSM):

- Language and ignorance trick us into thinking that basic biology is similar between sexes and genders.
- Female exclusion from basic biological studies creates a knowledge void that is carried through to clinical trials.
- GSM's origins in women's health and the word, gender, itself, continue to confuse people into thinking that the field is "really" about women's reproductive health.
- In most cases, the focus on biological mechanisms is taken to be antithetical to considering how the social world influences biology—or, how "the world writes on the body" [2].

In my career since 1995, my efforts have been focused on trying to bust these myths through writing, teaching, public service, and research. While there is always a vocabulary game—that is, putting these issues into language that someone who has never thought of them before can understand—persistence and persuasion through strong argumentation as well as teaching the next generation are still the most powerful tools we have. To quote Max Planck [3]:

> *"A new scientific truth does not triumph by convincing its opponents and making them see the light, but rather because its opponents eventually die, and a new generation grows up that is familiar with it."*

A JOURNEY

My concrete commitment to understanding sex and the biological body began in earnest when the Women's Studies Department at

Duke University asked me to be on their advisory board; they wanted a biologist to be involved in informing their curriculum. It was at that point, that I developed a course with substantive biology that could be cross-listed in both Women and Gender Studies and in Zoology, *Sex and the Brain*. I developed the course and ultimately a book of fundamental papers in the field of *Hormones and Behavior* which evolved into a discussion of sex differences, "Sex and the Brain" [4].

There is nothing like teaching a topic for learning a topic. One key class that raised the question I am still pondering was given by a faculty member in Women and Gender Studies whom I asked to speak to the difference between sex and gender. The gist of her answer in class was: "We don't think there is any." That is an answer and a tension between sex and gender I have chewed on and with which I continue to struggle: sex and gender are separable for purposes of doing an experiment but inseparable in lived lives.

Two points stood out for me after I had taught the course numerous times:

1. Humans do not show "sexual dimorphisms" of the brain and behavior in the way the rodents do—while there are statistically significant differences, the standard deviation is often larger in human than in rodent studies indicating enormous variability. I wondered whether life experience and human lives led to this variation in brain response and structures.
2. Even rodent sexual behavior is not as dimorphic as assumed. Early experiments by Beach and colleagues actually tried to show behavioral similarities rather than differences [4] and subsequent studies, in order to strengthen the effects of treatments or exposure, first eliminated animals from the study that did not perform with male-like or female-like behavior [4]. I thought that this complexity, even in nonhuman animal models, also pointed to variation in experience. This has been supported by the finding that the caging of male mice is a source of male behavioral variability [5].

In spite of the fact that I was trained as a neuroanatomist, this made me begin to question the biology-only explanation of sex differences. Couldn't experience, especially the social experience of being female or male (gender) mediate this variability? Since the groundbreaking work of Michael Merzenich we have known that experience modifies brain circuits [6].

THE WORLD WRITES ON THE BODY

In 2005, as an administrator of a women's health research institute, I was very perplexed about what women's health was and what it was becoming. It seemed that the fabric had many threads but no consistent weave. To help sort this out, my colleague, Margrit Shildrick and I organized an international "Think Tank to Retheorize Women's Health" which included invited participants from medicine, basic science research, social studies of science, and philosophy—faculty and graduate students—to consider how to realign women's health with current developments in biomedical technology, conceptions of the body, and feminist theory [7].

The mandate for participants was to transcend the constraints of their disciplinary affiliations and to identify and explore together the emergent topics that needed consideration for moving the principles of women's health forward. To generate discussion and raise critical points for collective consideration, five talks were given, each representing significant aspects of the theory, biology, and technologies of healthcare. These included: *Women's Health: Where We have Been/Where We are Going* (Adele Clarke); *Talking the Talk and Walking the Walk* (Gillian Einstein and Margrit Shildrick); *Race and Bones* (Anne Fausto-Sterling); *The Immune System: A Primer* (Michael Ratcliffe); *Tissue Engineering and Making Body Parts* (Kim Woodhouse). Including gender with sex emerged as a key theme of the think tank. One of the greatest pleasures was the "report back" by graduate students after which one of the faculty participants said, "I don't think we need to worry anymore about the future. The students can carry this forward."

This workshop and its subsequent paper highlighted some key strategies for forwarding GSM that are still important in today's climate:

- Multiple disciplines talking with each other and trying to connect with their deep similarities encourages innovation and new ideas.
- The larger question needs to be shaped to encompass ideas across disciplines.
- Engaging students in the process of wrestling with ideas creates new leaders.

This approach more broadly has enabled a move for women's health from being solely about reproductive health, to sex differences research, to incorporating sex and gender, that is, the biological and the social [7]. All have been critical perspectives from which to improve the health of women; the more recent inclusion of gender allows lessons learned from women's health to benefit the health of men as well. Social epidemiologists have long argued for the inclusion of sex and gender in thinking about health conditions as well as the etiology and progress of disease. In 2003, Nancy Krieger published a key paper explaining the concepts of sex and gender emphasizing that a disease etiology might be gender, or sex, or both [8].

I have carried this approach forward via my research program on female genital cutting [2,9], cognition in women with the breast cancer mutation gene with ovarian removal [10–12], and in cognition in transmen [13] as well as through program and course development. One such program, the Collaborative Specialization in Women's Health (University of Toronto: http://www.dlsph.utoronto.ca/programs/collaborative-specialization-in-womens-health/), was developed to engage graduate students in multidisciplinary perspectives on women's health including, sex and gender. Engaging students from Public Health, Psychology, Religion, English, and Information Science, to just name a few of the participating programs, multiple disciplines and perspectives inform each other through the program's diverse graduate students. The core course and the seminar series are organized to juxtapose a biological finding and theoretical papers for a given gendered health condition. For example, to understand a condition more common in women such as autoimmune disorders, we read a review paper on inflammation and the effects of estrogens on the immune system [14] and a chapter questioning the immune system model, linking it to larger societal metaphors [15]. In this way, we combine the body and the world and students come to understand the broader issues on which their disciplinary perspectives touch. Some important principles to include in cross-disciplinary research as I have practiced it are:

- Maintain disciplinary rigor but move beyond silos.
- Collaborate, since no one person can really "own" multiple approaches.
- Listen seriously to the other view to expand your own.

STRUGGLES AND SUPPORT

As I have worked to establish study sections that could review grant proposals on estrogenic action, programs that take a multidisciplinary view on women's health teaching both sex and gender, engaging support for research that takes both sex and gender into account, helping to develop sex and gender policy, and getting funding for my own research program, the biggest impediments have been "science as usual," or as Kuhn describes it, "normal science" [16], reviewers' lack of imagination, and seeing the problem only from one's disciplinary perspective. One helpful way to deal with this is to reinterpret sex and gender to fit the understanding of person or institution to whom one is speaking. For example, in health-care systems, taking sex and gender into account will help us understand some of the etiologies for chronic disease and the comparative approach will allow us to capitalize on the relative resilience of males and females.

Support for this kind of approach is rare. One's disciplinary colleagues do not always "get it" and often, it is not rewarded by an academic system that is organized into disciplinary silos. Since science is a process of building upon blocks of already existing knowledge [16], a new field is not likely to have as many blocks to build on in an established field. As well, the depth of knowledge in a multidisciplinary field may seem less than that in single disciplinary fields. Both of these issues make it difficult to obtain funding as well as garner institutional support.

There may be resistance to bringing the biological and social together based on a fear of complicating a research question by including females as well as males and considering gendered life experiences. Cahill points out that many scientists want to do "fundamental" research which should apply to everyone, not realizing that what is fundamental is human variation [17]. Research in GSM actually represents how complicated biology and health really are.

While there are luckily some faculty colleagues who will provide intellectual support for the approach, in general, academic institutions do not support multidisciplinary initiatives. A private donor specifically allocating monies for their support can, however, tip the scales. One such donor and fund raiser in Canada is Lynn Posluns, founder of the *Women's Brain Health Initiative* whose mission is to increase the knowledge about women's brain health and aging (https://womensbrainhealth.org).

Ms. Posluns has raised money to help support special multidisciplinary sex and gender research platforms in dementia research as well as established a chair in women's brain health and aging. There are certainly others around the world who have supported the multidisciplinary nature of GSM, providing monies for chairs and centers.

Certain countries may be more supportive than others. The notion that gender (the social) must be incorporated with the study of sex differences (the biological) is one that has fruitful ground in Canada and, indeed, some of the first researchers working on epigenetics (one of the mechanisms by which the world might enter the body) are Canadian [18,19]. In addition, with the reorganization of the Medical Research Counsel in 2000 into the Canadian Institutes of Health Research, Canada affirmed this position by making one of the 13 Institutes, The Institute of Gender and Health (IGH). The IGH offered funding opportunities to study sex and/gender as well as in training initiatives by IGH since its inception in 2000 [20]. The major success of these efforts has been in raising researchers' awareness that better science comes from considering sex and gender [21]. The IGH ensured that applicants for funding to CIHR address how they are or are not considering sex and/or gender in their applications, have begun to develop methods [22], training modules for researching sex and/or gender (http://www.cihr-irsc.gc.ca/e/49347.html), and developed training initiatives such as the Intersections of Mental Health Perspectives in Addictions Research [23]. During the 6 years I served on the IGH Advisory Board, I had the opportunity to help shape the direction of the IGH as well as sex and gender policy in Canada. Even with this central location, there were constant explanations of why sex and gender are important and how considering them leads to better science.

Some of the greatest support I have received has been from faculty in women and gender studies departments who recognize the importance of the biological body as well as of the social world: Jean O'Barr and Kathy Rudy (Duke), Margrit Shildrick and Nina Lykke (Linköping University), Afsahneh Najmabodi (Harvard University), Anne Fausto-Sterling (Brown University), and many others. I have always tried to keep one, small finger in that pie—both for the support but also because I think the theory can be instructive and lead to innovation in my own ideas for research. These colleagues introduced me to the works of Thomas Laqueur who showed that sex differences in the Western world were noticed due to the social needs of the 17th century [24]; Elizabeth Grosz who postulated a theory of a corporeal body that is both biological and social [25]; Nancy Tuana who founded a field of epistemology claiming that ignorance is as important as knowledge—particularly important in women's health [26]; and Emily Martin who did not question the scientific findings but rather reinterpreted them showing that gender can influence the interpretation of findings that actually showed something quite different [27]. These authors' theories have served as guideposts.

Support also comes from networks of colleagues around the world. My first involvement with like-minded colleagues was in 1995 when I was invited to a meeting of the Society for Women's Health Research (SWHR). This was also the first time I met Marianne Legato, the originator of GSM. Both, through their tremendous efforts to promote sex differences research and GSM have created courage and achievement paving the way for me and for many others. The Organization for the Study of Sex Differences and the International Society for GSM have been huge supports to the field both by promulgating the principles and science of this approach and providing an annual space in which those of us doing this kind of work can share our efforts, findings, and best practices.

CALL FOR ACTION

While the successes of the evolution of women's health into GSM are many, there are still obstacles to overcome:

- The conflation of "gender" with "women" in both the popular and the biomedical imagination.
- Resistance of scientists to consider the role gender plays in designing, analyzing, and interpreting their research—both human and animal models.
- The lack of a good measure of gender—all currently are woefully out of date.
- Practising physicians' moving away from considering the circumstances of their patients' lives due to the time constraints on treatment time.
- The conflation of the study of sex and gender differences with equity issues, which in some peoples' minds decouples GSM from the mandate to do rigorous science.

CONCLUSION

Recently, I attended the Canadian Association for Neuroscience meeting. When I asked one of the keynote speakers, an eminent scientist, if she had seen sex differences in her results she gave me the usual answer of "no." It was clear, however, that experiments were not designed to test or notice any differences. But at the posters, which are primarily presented by students, the answer was different. A few posters actually stated the sex of the animals studied. A few noted that they were designed to study sex differences, and others described how the study was powered to report sex differences. Even if the posters did not say the sex of the animals studied, when I asked, the presenter was sheepish, suggesting that they know this is an important question. Some respond that they just did not have enough animals or were only given males or females to study. A student said that she went to her advisor and said that there was a difference between male and female response because while she used the same brain coordinates to stimulate males and females, she thinks the equivalent area she wants to stimulate is represented in a slightly different brain location in females than in males.

One wonderful poster reported on how memory acquisition in snails (hermaphrodites) that originated in the same place in the Netherlands but which were being raised in different Canadian regions had different timings for learning depending on whether they "lived" in Ontario or Alberta, Canada. Snails in Ontario took longer to learn the same paradigm than snails in Alberta. When Alberta snails were tested in Ontario, their learning was slower and memory retention was not as good as when they were in Alberta; Ontarian snails taken to Alberta after a few sessions performed better than they did while in Ontario [28]. This student is studying how the world writes on the body—even the snail's. So, while we progress at a snail's pace, nevertheless, it is progress.

REFERENCES

[1] Boston Women's Health Book Collective. *Our bodies, ourselves*. 1st ed. Boston: Touchstone Books; 1976.

[2] Einstein G. Situated neuroscience: elucidating a biology of diversity. In: Bluhm R, Maibom H, Jacobson AJ, editors. Neurofeminism: issues at the intersection of feminist theory and cognitive science. 1st ed. New York: Palgrave McMillan; 2012. p. 145–74.

[3] Planck M. Scientific autobiography and other papers. 1st ed. trans. Gaynor F. New York: Philosophical Library; 1949, p. 22.

[4] Einstein G. Sex and the brain: a reader. Cambridge: MIT Press; 2007.

[5] Prendergast BJ, Onishi KG, Zucker I. Female mice liberated for inclusion in neuroscience and biomedical research. Neurosci Biobehav Rev 2014;40:1–5.

[6] Merzenich MM, Recanzone GH, Jenkins WM, Grajski KA. Adaptive mechanisms in cortical networks underlying cortical contributions to learning and nondeclarative memory. Cold Spring Harbor Symp Quant Biol 1990;55:873–87.

[7] Einstein G, Shildrick M. The postconventional body: re-theorizing women's health. Soc Sci Med 2009;69:293–300.

[8] Krieger N. Genders, sexes, and health: what are the connections—and why does it matter? Int J Epidemiol 2003;32:652–7.

[9] Einstein G. From body to brain: considering the neurobiological effects of female genital cutting. Persp Biol Med 2008;51:84–97.

[10] Einstein G, Au A, Klemensberg J, Shin EM, Pun N. The gendered ovary: whole body effects of oophorectomy. Can J Nurs Res 2012;44:7–17.

[11] Au A, Schwartz D, Eisen A, Finch A, Hampson E, May T, et al. Long-term effects of bilateral salpingo-oophorectomy on cognition. The Endocrine Society's 98th Annual Meeting & Expo, Boston April 1–4, 2016; abstracts in: Endocrine Reviews 2016;37 (poster LBFri-31). <http://press.endocrine.org/doi/abs/10.1210/endomeetings.2016.NP.4.LBFri-31>.

[12] Reuben R, Au A, Hampson E, Tierney M, Narod S, Bernadini M, et al. Changes in women's performance on the RAVLT over time post-oophorectomy. 11th Annual Canadian Neuroscience Meeting (Abstracts) 2017; May 28–31:223 (Poster 2-F-191).

[13] Watt S, Einstein G. Beyond the binary: the corporeal lives of trans individuals. In: Schreiber G, editor. Transsexualtität in Theologie und Neurowissenschaften. Berlin: Walter de Gruyter; 2016. p. 55–74.

[14] Fish E. The X-files in immunity: sex-based differences predispose immune responses. Nat Rev Immunol 2008;8:737–44.

[15] Haraway D. The biopolitics of postmodern bodies: determinations of self in immune system discourse. Differences: A J Fem Cult Studies 1989;1:3–43.

[16] Kuhn TS. The structure of scientific revolutions. 1st ed. Chicago: University of Chicago Press; 1962.

[17] Cahill L. Equal ≠ the same: sex differences in the human brain. Cerebrum 2014 (April);1–14.

[18] Weaver CG, Cervoni N, Champagne FA, D'Alessio AC, Sharm S, Seck JR, et al. Epigenetic programming by maternal behavior. Nat Neurosci 2004;7:847–54.

[19] Juster RP, Smith NG, Ouellet É, Sindi S, Lupien SJ. Sexual orientation and disclosure in relation to psychiatric symptoms, diurnal cortisol, and allostatic load. Psychosom Med 2013;75:103–16.

[20] Duchesne A, Tannenbaum C, Einstein G. Funding agency mechanisms to increase sex and gender analysis. The Lancet 2016;389:699.

[21] Johnson, JL, Greaves L, Repta R. Better science with sex and gender. <http://bccewh.bc.ca/wpcontent/uploads/2012/05/2007_BetterSciencewithSexandGenderPrimer-forHealthResearch.pdf>; 2007.

[22] Oliffe J, Greaves L, editors. Design and research in gender, sex, and health. 1st ed. San Francisco: Sage; 2012.

[23] Greaves L, Poole N, Boyle E, editors. Transforming addictions: gender, trauma, transdisciplinarity. 1st ed. New York: Routledge International; 2015.

[24] Laqueur Thomas. Making sex: body and gender from the Greeks to Freud. 1st ed. Cambridge: Harvard University Press; 1990.

[25] Grosz E. Volatile bodies: toward a corporeal feminism. 1st ed. Bloomington: Indiana University Press; 1994.

[26] Tuana N. The speculum of ignorance: the women's health movement and epistemologies of ignorance. Hypatia Summer 2006;21:1–19.

[27] Martin E. The egg and the sperm: how science has constructed a romance based on stereotypical male-female roles. Signs 1991;16:485–501.

[28] Rothwell C, Spencer G, Lukowiak K. The influence of environmental factors on memory formation. 11th Annual Canadian Association for Neuroscience Meeting (Abstracts) 2017; May 28–31:109 (Poster1-F-182).

CHAPTER 7

The Charité Approach—Sex and Gender in Research, Teaching, and Policies

Vera Regitz-Zagrosek
Institute for Gender in Medicine and Center for Cardiovascular Research, Charité, University Medicine Berlin DZHK, partner site Berlin

The Institute of Gender in Medicine, the only one in Germany, and one of the three in Europe, was established at Charité University Medical School in 2003 as a "loose connection of researchers and clinicians" led by Vera Regitz-Zagrosek, MD, before becoming a formal Charité institute in 2007. It received early support from the Berlin gender equality program, German Heart Centre Berlin, Humboldt University, and Free University of Berlin. By 2004 it had attracted 80 associated members from 21 Charité institutes. All major medical disciplines were represented, and close cooperation with social sciences was established. However, the focus was cardiology and cardiovascular research, reaching from basic to clinical studies and public health. The overall goal was and is improving the treatment of sex and gender differences in frequent diseases the health care system for women and men, and the consideration of sex and gender aspects in public health needs adequate training of all parties involved, medical students, other students, professionals, doctors and researchers, health politicians.

FROM A STUDENT TO PROFESSOR IN GENDER MEDICINE

Professor Regitz-Zagrosek decided on a research career after participating in the "Jugend forscht" in physics while at high school, where she earned a stipend from the Studienstiftung des Deutschen Volkes and met researchers from the Max Planck Society. After medical school she received her postdoc training at the Max-Planck-Institute for Experimental Cardiology and at University of Madison, Wisconsin, Department of Biochemistry.

The International Society for Gender Medicine. DOI: http://dx.doi.org/10.1016/B978-0-12-811850-4.00007-7

From 1981 to 1984, Professor Regitz-Zagrosek was working as physician scientist, clinical cardiologist, and finally responsible for the outpatient department of Deutsches Herzzentrum Berlin from 1985 to 2003. With funding from the Deutsche Forschungsgemeinschaft and some institutional support, she was able to continue her experimental work on myocardial energy metabolism and became involved in clinical studies. She then became associate professor at the Free University Berlin in 1993, and from 1996 to 2001, she was head of the Outpatient Department of Cardiology. In this clinical role, Professor Regitz-Zagrosek carried out >60 hours of clinical work a week and saw >3000 patients a year. She recalls experiencing a "strict clinical hierarchy" in the Department of Cardiology at the German Heart Institute in Munich and later at Berlin that opened her my eyes to the imbalance between women and men as patients and doctors in cardiology. From there, she became interested in gender medicine.

From 2001 to 2010 she founded and coordinated the Deutsche Forschungsgemeinschaft-funded interdisciplinary graduate course on sex differences in myocardial hypertrophy, which helped to train >50 students and brought an overwhelming richness of ideas and input from different participants. The course disseminated the ideas of gender research among Charite researchers. As a consequence she could continue the experimental research on a larger basis, in the DFG Research group on "Sex-specific mechanisms of myocardial hypertrophy" (2008–14).

In 2002, she founded a working group on "Cardiovascular Disease in Women" at the German Cardiac Society and chaired it for 6 years. This group created awareness that "women are not just small men" among German cardiologists.

In 2003, she obtained a chair on Cardiovascular Disease in women at Charité Berlin. She founded the Institute of Gender in Medicine (GiM) at Charité in 2003 that developed from a loosely organized center into a successful institute. In 2007 the founders received the Margherita von Brentano Prize of Free University of Berlin. Clinical studies and public health-related activities account for ≈30% of its work, whereas ≈70% of the activity is basic research. The Institute of Gender in Medicine is proud of its contribution to a new area in cardiovascular research by focusing on sex and gender differences in different animal models and clinical studies. The multinational team, with VRZ as a director, consists of up to 10 postdocs, 5 technicians, 2 project assistants, usually ≈ 10 MD and PhD students, and public health master students. She also founded the German and International Society for Gender in Medicine in 2007. International congresses were organized, in Berlin, Stockholm, Tel Aviv, again in Berlin.

Professor Regitz-Zagrosek published over 200 scientific papers in excellent journals and numerous book chapters, edited 2 landmark books on Gender medicine and organizes biannual international congresses on Gender Medicine. She acts as reviewer for national and international funding organizations and journals. In 2015, she received the Honorary doctorate of the Medical University of Innsbruck and in 2016, she had the opportunity to lecture at Academie Francaise in French on Gender Medicine.

GENDER IN RESEARCH

Among the most important basic research carried out by the Institute of Gender in Medicine team is that on cardiac functions of sex hormones and the effects and mechanisms of sex-specific genes in cardiovascular tissues [1,2]. In the human heart, estrogen activates different pathways and induces different genetic programs in males and females. In male human hearts, estrogen activates some pathways that lead to an impairment of contractility in males only. This may lead to new therapeutics in women and men. Being able to dissect sex-specific pathophysiological mechanisms in male and female hearts has been the most exciting recent development. This includes a number of protective pathways that are unique to the female hearts, not necessarily due to direct sex hormone effects. Genetic imprinting and epigenetic modifications differ in male and female hearts and may significantly contribute to sex-specific responses. In clinical studies, much progress has been made to delineate sex- and gender-specific mechanisms in human diseases [3].

Funding for gender research was first obtained at European level. EUGIM (2008–10) developed an European curriculum in Gender Medicine. EUGENMED (European Gender Medicine, 2013–16) aimed at establishing a roadmap for implementation gender into European biomedical research and GENCAD "Gender in coronary artery disease" (2014–17) aims at collecting facts, knowledge, and

Figure 7.1 The founding members of the German Center for Cardiovascular Research, Berlin 2011.

awareness on sex and gender in coronary artery disease. VGM also coordinated an EU training network in basic science (RADOX), introducing sex and estrogen effects there. The European Society of Cardiology (ESC) also established a Task Force for the Guidelines "Cardiovascular Diseases in Pregnancy" and she became task force leader for 2010–12 for the initial Guidelines and in 2016 for their revision.

Funding for sex and gender topics in medicine in Germany was rather poor and she stayed partially in the cardiology world, slowly introducing gender topics there. Now, she coordinates the Berlin site in the "German Centre for Cardiovascular Research" (DZHK, BMBF, 2011–18), and in the German DZHK she remains the only women in a male environment (Fig. 7.1). Focus of her work are sex- and gender-specific mechanisms in heart failure. Constant lobbying helped to install some gender projects in Germany—among them the pilot project "Gender Medicine" (BMBF) where a large database was generated, collecting and classifying all pubmed publiations in gender medicine. This led the scientific ground for two textbooks [4,5]. Finally a large project GENDAGE (2017–21) will be funded by BMBF, dealing with gender and age in the Berlin and German population.

TEACHING

The GIM developed new concepts in teaching Gender medicine as coordinator of the European project EUGIM (European Curriculum in Gender Medicine) and implemented them at Charité.

Medical Students

We established at Charité Universitätsmedizin Berlin a curriculum on Gender Medicine that is now implemented into the medical student's regular curriculum. It starts with an introductory lection in the first semester in the very first days of their study and continues throughout the studies. Wherever possible, sex- and gender-related issues are integrated into the specific disciplines, that is, into the cardiovascular curriculum, into psychiatry or surgery or many others, in more than 100 seminars, lectures, and trainings. However, in Module 35, in the 4th year, sex- and gender-specific diseases, the students are offered courses in core knowledge areas and competences of gender medicine; that is,

- sex and gender differences in mechanisms of disease
- sex and gender differences in pharmacology
- sex and gender differences in clinical manifestations and outcomes of frequent diseases.

Sex is also integrated in the basic research areas where the effect of chromosomal regulation and hormones is discussed. In all parts, sex and gender are elements of examinations and written questions, multiple choice as well as others have been elaborated.

In addition, we established an elective course that gives interested students the opportunity to learn more about sex and gender medicine in one single course and to get an overview on this evolving area.

Knowledge basis: in order to establish a reproducible and reliable knowledge basis, we established a database on gender medicine literature, where we have now assembled and classified and analyzed more than 5000 publications in the field. This was made possible by 6 years funding of BMBF. The database (www.gendermed.db) is accessible from outside and helps to search classified and quality-proven gender medicine papers according to topics, authors, and more criteria. Using this database avoids the fighting with the many false positive papers that appear with searches in public databases and it helps getting a structural overview on the available references in a given field and the type of information they offer.

Nonmedical Students and Professionals

We developed two modules for master programs that can be integrated in differing master programs, for example, there is now a module Gender Medicine in the Berlin School of Public Health and there is a module integrated into Health and Society. In these courses mainly public health students that will go into health care management, assurance companies or political organizations will be exposed to Gender

Medicine. The elective modules are very well accepted. They are now taught for the second time in 2012. In order to reach as many professionals as possible we established an e-learning course on Gender Medicine where the most important messages of this field are communicated into the scientific community. So far it is used in combination with 2 days of personal tuition and a personal examination, that is, in a blended learning program.

Teaching Book

We have also established and edited a teaching book on Gender Medicine where the most important fields of knowledge are expanded. This teaching book is comprehensive focusing on clinical questions, and offers a large number of illustrations and overview tables. It is based on a systematic literature search where more than 10,000 references have been included and categorized according to epidemiology, pathophysiology, clinical manifestations of disease, management, and outcome. This will certainly contribute to set the stage in Gender Medicine and to define the available knowledge and the scope of knowledge.

Qualification in Gender Medicine

To improve knowledge of gender specific medicine in the medical profession the German Society of Gender Medicine (DGesGM e.V.) has set up a qualification as Gendermediziner DGesGM. Only physicians can acquire this qualification. The qualification Gendermediziner makes clear that physicians have been exposed to the goals and principles and knowledge of Gender Medicine in a way, that she/he is able to treat women and men in a gender-sensitive manner. Documented activities in the field of Gender Medicine as well as participation in training courses are required to obtain this additional qualification.

Curriculum in Gender Medicine

The Institute of Gender in Medicine (GiM) at Charité Universitätsmedizin Berlin also coordinated a European project (EUGIM) from 2009 to 2011 that defines a curriculum and study regulations in gender medicine. It follows the principles of the Bologna process that aims at harmonizing curricula between European universities. It therefore established learning goals, skills, and required investment of time in a way that this curriculum can be taught throughout Europe. So far we have packed it into two modules for master programs that can be integrated into diverse biomedical or social sciences master program.

REFERENCES

[1] Regitz-Zagrosek V. Therapeutic implications of the gender-specific aspects of cardiovascular disease. Nat Rev Drug Discov 2006;5(5):425–38.

[2] Regitz-Zagrosek V, Kararigas G. Mechanistic pathways of sex differences in cardiovascular disease. Physiol Rev 2017;97(1):1–37.

[3] EUGenMed Cardiovascular Clinical Study Group, Regitz-Zagrosek V, Oertelt-Prigione S, Prescott E, Franconi F, Gerdts E, et al. Gender in cardiovascular diseases: impact on clinical manifestations, management, and outcomes. Eur Heart J 2016;37(1):24–34.

[4] Oertelt-Prigione S, Regitz-Zagrosek V. Sex and gender aspects in clinical medicine. London: Springer Verlag; 2011. p. 201.

[5] Regitz-Zagrosek V. Sex and gender differences in pharmacology. Handbook of experimental pharmacology. Heidelberg: Springer Verlag; 2012. p. 600.

CHAPTER 8

Gender-Specific Medicine in Italy: Point of View and Journey of Giovannella Baggio

Giovannella Baggio
University of Padua, Padova, Italy

When you do something,
you have to know that you will have against you people who liked to do the same thing,
People who liked to do the contrary,
And the vast majority of People who liked to do nothing.

Confucius

INTRODUCTION: PERSONAL ITINERARY

I graduated in 1972 and in the early 1980s I began to be interested on the differences of risk factors for cardiovascular diseases in women. It was not easy to convince even scientific medical journals about the importance of this dimension of medicine, so that I had to publish my observations about an important topic as "The difference of postprandial phase in postmenopausal and fertile women" in a Gynecological journal!! [1]. Meanwhile, in the 1980s the Lorenzini Foundation (Milan-Houston) was addressing the international community about the importance of studying the differences in medicine for women. The Foundation organized several conferences in Italy and United States: I was very much involved in these conferences as well as their publications. In the 1990s my interest increased. At that time I was Full Professor of Geriatrics and Gerontology in Sassari, Sardinia Island, and I started to work with the Sardinian Centenarians, focusing the attention on extreme longevity in men and women.

In 2005 I was motivated to pursue Gender Medicine (GM) by an inspired gynecologist who told me "there is a new and very interesting field to which I cannot contribute because I work only with women: Gender Medicine!" And with her I attended to the first International

The International Society for Gender Medicine. DOI: http://dx.doi.org/10.1016/B978-0-12-811850-4.00008-9

Gender Medicine (IGM) Congress in Berlin in 2006, where I met for the first time Marianne Legato, the founder of GM in the World and one of the founders of the International Society for Gender-Specific Medicine (IGM). For the first time in Europe, during the IGM Congress it was proclaimed that GM is not concerned only with women's health or women's diseases, but with the differences between men and women's experience of disease (cardiovascular, kidney, oncology, etc.): the daily work of a doctor!

In 2009 with a group of friends (high-level professionals in medical science: MDs, University Professors, Health Managing Experts together with the collaboration of University Hospital of Padua and the Lorenzini Foundation Milan), we decided to found the Italian Research Center for Gender Health and Medicine (*Centro Studi Italiano per la Salute e la Medicina di Genere*, CSSMG). This Society very quickly attracted numerous members from all over Italy and was recognized as a member of the International Gender Medicine Society. In these past 8 years, we concentrated intensely on implementing education, information, research, and the amplification of the Italian network. I was involved firsthand in all of these activities: my personal history with GM can be identified with the activities of the CSSMG.

Educational Activities

We have done and/or participated in 40 events per year throughout Italy on GM. In 2009, 2011, and 2013 we organized the first, second, and third Italian Congresses on GM. In 2011 it was held the first Congress on Gender Oncology (perhaps first and unique in the World). In 2015 we participated in the International Berlin IGM Congress (with 30 posters), and this year (2017) we are organizing the fourth Italian Congress on Gender-Specific Medicine.

Information Activities

In collaboration with many voluntary associations, we participated in many events for the lay public. I would like to mention the more active associations: AMMI (*Associazione Mogli e Mariti di Medici Italiani*: Association of Wives and Husbands of Italian Medical Doctors), Soroptimist, Fidapa, some Labor Unions, Italian Association of Female Doctors (AMD), Women and Science (*Donne e Scienza*), the Italian League against Tumors (*Lega Italiana contro I Tumori*, LILT),

Equal Opportunity Groups committees. In addition, the AMMI announces every year a scholarship competition (10,000 Euro) for a young doctors who will pursue a research project in GM.

Research

The numerous stimuli to all fields of medicine and the collaboration with different scientific societies significantly increased thc number of scientific publications by Italian investigators, as can be noted in PubMed. I would like to remember the particular important contributions of Walter Malorni, Claudio Franceschi, Elena Ortona, Marina Ziche, Daniela Monti, Flavia Franconi, myself, and many others.

Creation of a Network

This was the most difficult task, but we are very proud of our success. Walter Malorni (Chief of the National Center for Gender-Specific Medicine of the Italian National Institute of Health, ISS), Annamaria Moretti (President of the Italian Group of Gender Health, GISeG, *Gruppo Italiano Salute e Genere*), and myself (President of the CSSMG) united in a strong alliance and we become "the carrier trio" of the Italian network (Fig. 8.1). As you can see in the figure, CSSMG forms the connection with IGM. The Lorezini Foundation is very collaborative and a cofounder of CSSMG. FNOMCeO (the Federation of all Italian General Medical Council, consisting of 400,000 physicians) has created a Committee on Gender Medicine, and the Italian Society of General Medicine Doctors (SIMG), composed of 45,000 physicians, has many activities in GM. The FADOI (Italian Federation of Hospital Internal Medicine Specialists) has many educational and research programs. In addition, many of Italian Scientific Societies have a subsection on GM. Recently the Coordination Committee of the Italian University Schools of Medicine has voted for a program for teaching gender differences in all specialties in the Medical Schools. Finally many political regions put GM in their political programs. The Italian Network is crowded, diverse, and very active.

It does not matter how slowly you go so long as you do not stop.

Confucius

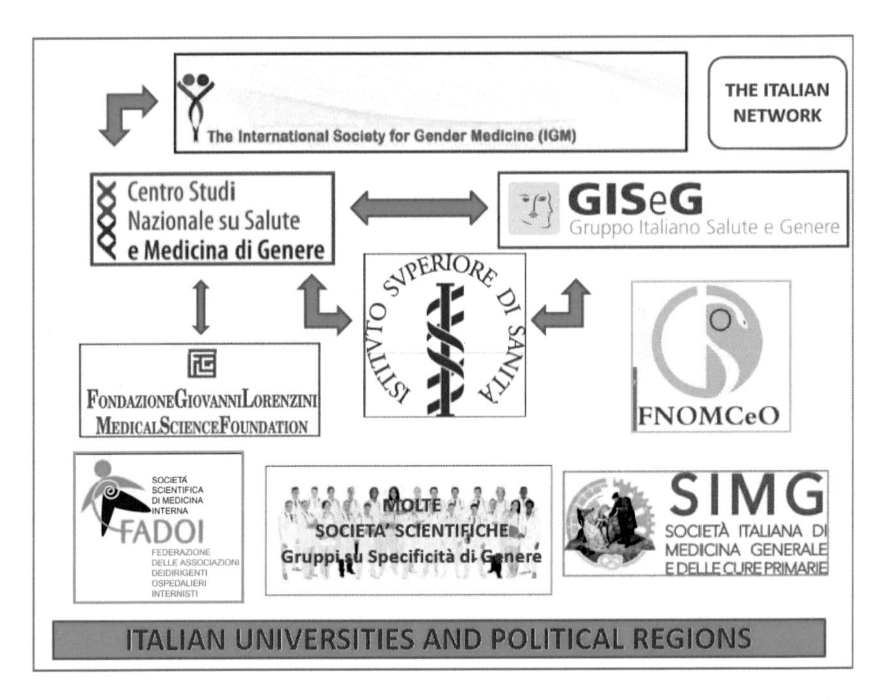

Figure 8.1 The Italian Network: was difficult to create, but is a very satisfying achievement. Modified from Il genere come determinante di Salute, Quaderni del Ministero della Salute, 2016, 26: 87.

Here below all societies of the Italian Network are listed, each with a brief presentation.

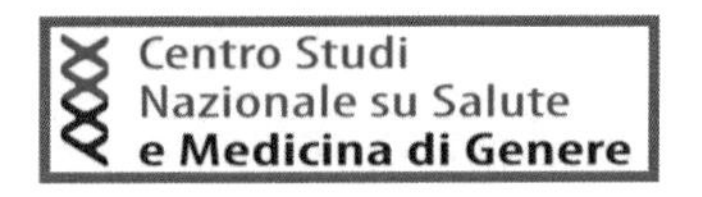

The National Research Center for Gender Health and Medicine (Italian CSSMG). President: Giovannella Baggio. www.gendermedicine.org

The National Research Center for Gender Health and Medicine (Italian CSSMG) was founded in 2009 by well-known experts in basic and clinical sciences and management, together with the University Hospital of Padua and the Giovanni Lorenzini Medical Science Foundation (Milan-Houston).

The objective of the National Research Center is to unite medical experts in producing increased information, education, research, and clinical care linked to the gender differences (see earlier).

The Italian CSSMG was very much engaged in creating the Italian Network on GM. It also has the role of stimulating and facilitating the acceptance by all health professionals of GM that, unfortunately, have been for too long neglected not only in both basic and clinical research but also in everyday clinical practice.

The Center of National Health and Gender Medicine is a member of the International Society of Gender Medicine, and the President of the Center sits on the board of that organization.

National Center for Gender-Specific Medicine of Italian Institute of Health (ISS). President: Walter Malorni. http://newsletter.iss.it/centro-di-riferimento-medicina-di-genere/

After many years of work carried out by a section devoted to the study of GM within the Department of Drug Research, an autonomous department named the National Center for Gender-Specific Medicine (MEGE) has recently been created (January 1, 2017) at the Italian National Institute of Health (*Istituto Superiore di Sanita*, ISS). The head of the Center is Dr. Walter Malorni, involved with his research group for many years in studying the biology of sex differences. The activity of the Center is focused on three main objectives: (1) to promote research of identifying the pathophysiological bases responsible for gender and sex differences; (2) to develop training and educational activities dedicated to GM; (3) to develop a network of Italian Centers and laboratories that deal with GM. In addition, in view of the role played by the ISS as a technical-scientific body of the Ministry of Health, the Center will have the task of linking the various institutional stakeholders already operating in the field of gender-specific medicine, and promoting this new approach within the National Health System (SSN). These activities are already ongoing thanks to fruitful collaborations with the Italian Parliament, universities, medical and patient associations, Italian Regions, and agencies involved in patient's cures such AIFA (the Italian agency in charge of drugs and clinical trials). The structure of the MEGE is composed by more than 50 individuals (researchers, technicians, and fellows) who are divided into Operating Units, which have the task of carrying out institutional research, educational and training activities. These are aimed at improving the preventive, diagnostic, prognostic, and therapeutic strategies of all illnesses that show differences between the two sexes. In particular, gender-specific diagnostic and prognostic biomarkers, oncology, autoimmunity, nutrition and lifestyles, cardiovascular disease, preventive medicine, and toxicology represent some of the fields of interest and competence. The major competence of the MEGE is in the fields of gender pharmacology and cytopathology. The role of sex hormones and genetic and epigenetic factors in the pathogenesis of certain diseases such as tumors and immune-mediated, cardiovascular, and metabolic diseases are also being studied. Finally, gender differences in response to specific dietary, genetic, metabolic, and behavioral regimens are also investigated. Currently, particular attention is given to infectious diseases, in particular in the responses to antibiotic treatment and vaccines.

Italian Group for Health and Gender. President: Annamaria Moretti. www.giseg.it

The Italian Group for Health and Gender (GISeG) was founded in Bari in 2009, with the aim of implementing strategies for the promotion of a gender health culture through primary and secondary prevention programs. The Group includes interactions with hospital, institutions, GM doctors, professional boards, nonmedical health specialties, and patient associations. The aim of the Group is to provide: correct information on gender differences, definitions of preventive care paths, and the development of professional training activities. Over the years, GISeG has created its own website, organized an annual National Congress, and has participated in National and International Congresses on GM. Currently, the activities within Puglia Region of epidemiologic studies aim to collect data on "Respiratory Diseases and

Gender Differences." The Group participates in regional and national boards (General Medical Council, FNOMCeO), and the scientific board of the *Italian Journal of Gender-Specific Medicine* and has contributed with many scientific publications.

The Giovanni Lorenzini Medical Science Foundation in Gender Medicine. President: Andrea Peracino; CEO: Emanuela Folco. www.lorenzinifoundation.org

In 1988 the World Health Organization has issued a "gender challenge" to national and international organizations, a call for: a better appreciation of risk factors involving women's health; the development of preventive strategies to lessen the impact of diseases that disproportionately plague older women (e.g., coronary heart disease, osteoporosis, and dementia); and an increased emphasis on understanding why men die sooner than women [2]. From that time, programs and projects on the gender challenge in different disease states, including atherosclerosis, cognitive impairment, and dementia are started.

In the 2010 the Giovanni Lorenzini Medical Science Foundation decided to concentrate on GM and established the European Society of Gender Health and Medicine.

The aims of Lorenzini Foundation in the field of GM, to mention a few, are:

- *to link* all the scientists and physicians who operate in the field of prevention, primary, and secondary care, and rehabilitation by focusing on the biological, physiological, and pathological differences between women and men;
- *to support* researchers, medical doctors, institutions, and individuals to identify healthcare issues and protect health of both women and men;
- *to improve* the cultural background, professional education, and the training of experts in GM;
- *to promote* the inclusion of gender perspectives in the programs of both public and governmental institutions;

- *to develop* alliances with research centers, scientific societies, hospitals, academia;
- *to realize* personalized medicine; and
- *to educate* the public on the gender differences in healthcare needs.

As of May 24, 2017, more than 90 projects as activities, or congresses, or events on GM, have been organized by the Lorenzini Foundation in, and outside of, Italy.

Italian Federation of Medical Councils. President: Roberta Chersevani; President of the Gender Medicine Commission: Teresita Mazzei. https://portale.fnomceo.it/

In 2015 the Italian National Medical Council that is the Federation of 106 provincial medical councils distributed in all the Italian area (FNOMCeO) established a Gender Medicine Commission. The Commission in turn is composed of four representatives of medical societies with expertise in this field (the Italian Society of General Medicine, SIMG), the Italian Group of Health and Gender (GISeG), the Federation of Associations of Hospital Internal Medicine (FADOI), the Full Professor of Gender Medicine at the University of Padua and President of the Italian Research Center for Gender Health and Medicine (Prof. Giovannella Baggio), and the Director of the Center of Gender Medicine of the Italian National Health Institute (Dr. Walter Malorni). They are supported by professionals belonging to various Provincial Medical Councils, and their activities are coordinated by Prof. Teresita Mazzei (Full Professor of Pharmacology at Florence University). The main objectives of the Gender Medicine FNOMCeO Commission are: continuing medical education (CME), spreading of information, contribution to a National Network, cooperation with political institutions, and stimulation of clinical research.

The Continuing Medical Education consists of the organization of CME courses at the various sites of Provincial Councils involving the presence of at least one member of the Commission. In 2016, more than 10 courses have been organized in the south and center of Italy with great success. The planned 2017 CME events started in February and have been scheduled mainly in the north of Italy.

FNOMCeO is also very interested in Distance Learning and Continuing Medical Education. To reach this goal, a Working Group, coordinated by Giovannella Baggio, has been recently established.

FADOI (Italian Federation of Hospital Internal Medicine Specialists) and Gender Medicine. President of Gender Medicine Commission: Cecilia Politi. https://www.fadoi.org/

The FADOI pathway in promoting GM for a cultural and educational process, which involves more than 2500 members, began in 2008. Since then, a Gender Medicine Area has been created, which promoted the GM annual course at the FADOI National Congress and the presentation of GM scientific reports at the FADOI Regional Congress, beginning and pursuing educational in GM for Italian hospital internal medicine specialists.

A gender section has also been introduced to the *Italian Journal of Medicine* (the official FADOI journal) with the publication of articles such as the post-hoc analysis by gender of ATA-AF FADOI-AMCO Study on Atrial Fibrillation (with more than 7000 enrolled patients).

In recent years, more than 50 FADOI speakers, experts in the major branches of Internal Medicine, have spread the science of GM not only in FADOI National and Regional Congresses, but also in national and international non-FADOI events. In 2015 the young FADOI (members <40 years), from various Italian regions, divided by areas of interest and in study groups, entered the arena of GM. This group worked for the 2016 GM course "Juniors meet seniors," and at 2017 National Congress they organized a very interesting course on "Internal Medical Complications of Pregnancy." The young FADOI GM group has also reviewed *by gender* two FADOI clinical trials: the FADOI DIAMOND on diabetes (>3500 enrolled patients) and FADOI DOMINO on pain (over 5200 medical charts examined). Both papers are in press. They have also submitted a series of articles for the *Italian Journal of Medicine* on gender differences in hepatitis, thromboembolism, CV risk, etc.

FADOI, the full membership of the association for GM, has recently promoted a common platform on gender culture with the Federazione Italiana dell Società Mediche (FISM) with three objectives: education, information, and appropriateness. FISM, the federation representing all the Italian scientific societies, plays a key role not only in shaping and sensitizing healthcare professionals to this new approach to medicine, but also in spreading guidelines and promoting proactive behaviors on the part of patients.

Italian College of General Practitioners. President of the Gender Medicine Commission: Raffaella Michieli. https://www.simg.it/

In 2007 the Gender Medicine Commission of the Italian College of General Practitioners published a Decalogue committed to highlighting and improving the knowledge in the field of GM by: stimulating gender research in all fields, activating collaborations for gender research, emphasizing gender issues in research (sample selection, side effects, etc.), identifying evidence of gender issues in the literature (better/worse effect of a drug, preventive action or diagnostics, etc.), implementing public promotion researches in gender differences through the website and the journal, dealing with gender-specific issues in all congresses organized by SIMG, addressing gender issues in the journal SIMG, paying greater attention toward suspicious and possible situations of violence against women, and supporting greater attention to suspicious and/or confirmed gender discrimination.

The society recognizes the need to balance the presence of physicians of both sexes in all social roles. Furthermore, we wish to urge the scientific secretariat to encourage an active collaboration of all the colleagues of Italian College of General Practitioners to ensure an appropriate representation in the Society among women and men. Several GM-related papers have been published as: "Mind, Heart, Arms and ... Women's Health Guide" by Elvira Reale, "Gender Medicine, a new frontier in medicine" by Ed. Hyppocrates, "Best friends" by Giunti Editore Publisher, and "Cardiovascular risk and gender differences" by SIMG Disease Management. The College attended in 2007 Ministerial Advisory Council of the Women's Health Board and it is part of the Study, Research, and Documentation of Gender Medicine of the FNOMCeO 2014 working group. SIMG inserted two chapters on the topic of GM in the journal of SIMG and talked about it constantly within congressional sessions both at the regional level and during the annual national congress. Gender-based researches in the fields of heart failure and chronic obstructive pulmonary disease have been done. Moreover, an impressive poster on Violence on Women has been on display in all the waiting rooms of General Practitioners' outpatients, along with a questionnaire dedicated to women. Training courses for doctors on the topic of violence were constructed. These were all efforts of the project on violence against women named "VIOLA."

ITALIAN UNIVERSITIES AND POLITICAL REGIONS

Italian Universities

The Permanent Conference of the Presidents of the Italian Universities Medical Schools approved a document stating the need for teaching gender differences in all medical subspecialties. This is a very important stimulus to highlight the need for a separate Chair on Gender Medicine.

The Chair of Gender Medicine at University of Padua: A Start-Up

In 2013 the Department of Molecular Medicine of the University of Padua (a Department which includes all disciplines from the preclinical sciences to translational medicine) decided to found a Chair of Gender Medicine and nominated Giovannella Baggio (already full professor of Medicine) as person of the highest repute for this Chair.

Thus, since 2013–14 academic year, the Medical School of Padua has a course on GM. This course, however, is optional mainly because the Presidency of the School (as well as lot of University Professors) has difficulties in understanding the significance of GM. However numerous students attended to this course and were gratified to learn about this new field. The course of GM is organized as a platform on which the most famous professors of different specialties of Padua Medical School were appointed to give lectures on gender differences in his/her particular specialty. The course has three aims: teaching students, increasing awareness of professors (and pushing them to include gender differences in their own courses), and making students feeling responsible for knowing the importance of gender in all the specialties of medicine.

This point of consciousness of gender differences in medicine is in an extremely important first step: a kind of *start-up*!

COMMUNICATION TOOLS WITHIN AND OUTSIDE THE NETWORK

The Newsletter

In 2016 the ISS MEGE, the Italian CSSMG and GISeG, succeeded in unifying the newsletter: this was a strategic achievement to reinforce the network. It is published quarterly and is the perfect evidence of collaboration within the Gender Medicine Italian network.

It is disseminated *via* web and sent to 15,000 email addresses. http://www.lorenzinifoundation.org/newsletter-sulla-medicina-di-genere/

The *Italian Journal of Gender-Specific Medicine*

A flagship/crown jewel of the Italian Network is the *Italian Journal of Gender-Specific Medicine* founded in 2015. The *Italian Journal of Gender-Specific Medicine* is the first European scientific journal dedicated to GM, published by *Il Pensiero Scientifico Publisher* with the contribution of Novartis and the scientific support of the Editors and of the Editorial Board composed by the major Italian experts in the field. The *Italian Journal of Gender-Specific Medicine* is a quarterly journal that aims to make a contribution to the development and diffusion of a gender-specific medicine between clinicians, researchers, and decision makers, but also among the lay public, with a website that can be consulted by everyone: www.gendermedjournal.it

The publication's mission is to promote the field of GM including basic, clinical, and translational research in the field of biomedical sciences as well as in the fields of sociology and psychology. Policy and epidemiology also fall within the scope of the Journal. Moreover, the Journal publishes news, interviews, and reviews dealing with the role of sex and gender in the management of patients worldwide. In particular, there are three main types of contributions: "Review articles and Original articles," "Gender-specific medicine watch" (a summary of particularly important contributions issued in the international journals, announcements, and reports of events and conferences, etc.), "News and Perspectives" (news, reports, interviews with the main characters of the institutional and regional healthcare system, reports of legislation, and updates on European, national, and local regulations). In other words, the publication's mission is to promote the field of gender-specific medicine in the broadest sense.

Italian Political Dimension

In the Italian Parliament three legislative proposals were submitted in the last 3 years, but the Health Commission did not yet examine them in order to bring the proposals to be approved into the plenary assembly. However an amendment has been introduced in a forthcoming law on professionals' health workers, which includes all the issues of the proposed laws.

This will be a great step forward to improving GM in Italy.

> *The first woman was created from the rib of a man. She was not made from his head to top him nor from his feet to be trampled by him, but from his side to be equal to him.*
>
> ***Confucius***

CONCLUSION

I have pursed gender medicine in the path of my life over the last 10 years with compelling enthusiasm. I have made a great effort to construct an Italian network for the promotion of gender-specific medicine; to create bridges between individuals and organizations and to stimulate the interest of involved persons and institutions. This challenge was difficult for two reasons:

- as a Chief of the Internal Medicine Unit of the University Hospital of Padua, I have many clinical, teaching, and administrative duties.
- making bridges is often difficult because of jealousy, nevertheless, Gender-Specific Medicine is for me a fascinating field of work and a scientific clinical duty.

The network we have developed in Italy is a fantastic achievement through which many friendships have been formed: which we had not predicted!

REFERENCES

[1] Baggio G, Gabelli C, Fellin R, Martini S, Andrisani A, Baldo Enzi G, et al. Post-prandial behaviour of lipoprotein and apoprotein levels in post-menopausal and young fertile females. J Gynaec Endocrin 1986;I(3–4):51–7.

[2] World Health Organization. The World Health Report 1998, Geneva. <http://www.who.int/whr/1998/en/>; 1998.

CHAPTER 9

The Future of Science Through the Lens of Gender-Specific Medicine: Rewriting the Contract

Giuseppe Caracciolo and Alessandro Casini
Menarini International Foundation, Florence, Italy

The Menarini International Foundation was founded 40 years ago in Florence, Italy. We have preserved the original and important philanthropic roots of the Foundation in science, medical education, and culture. Over the last 5 years, we have built a powerful, diverse international presence. Making the difference nowadays is a must, even though it is not easy at all!

When we decided to improve the sophistication and administration of our Foundation, we faced triumph and tribulations. We risked failure in changing what had already been attempted by others. We carefully evaluated all our different possibilities, and considered the pros and cons for every single road we decided to follow. It took a little while before deploying our international activities, because a wrong step forward could have impacted our long history of important scientific contribution and excellence. Courage, enthusiasm, and passion have driven us to confront risk.

At the beginning of 2013, we initiated a new phase: moving our programs to the international level. While we knew that the opportunities and attractive goals were there, we also realized that in all honesty, the challenge was incredibly high. The social, economic, cultural, and educational societies and communities are unique in each country. Ultimately, we decided to develop the highest quality of educational programs possible in medicine, science, and culture, placing particular emphasis on medicine. We want to improve scientific knowledge at every level of expertise and training starting with students, and progressing to the most expert opinion leaders in the field. We were

The International Society for Gender Medicine. DOI: http://dx.doi.org/10.1016/B978-0-12-811850-4.00009-0

determined to achieve all this as a nonprofit organization. We were particularly interested in having a presence in emerging countries, where access to education is expensive and difficult. With the collaboration of the best recognized experts in each medical field, we reach not only the biomedical community but also patients and all others interested in the topics we cover. It is a great privilege and honor to make these programs available, which, if well used, will increase the scientific and cultural education for generations, making a real contribution and difference to the world's diverse societies. We believe that our contribution will help build a better world, less encumbered with inequality and discrimination.

Medicine has dramatically improved over the last decades [1]. A great deal of research over the past few years has provided us with previously unimagined descriptions of the physiologic and pathological mechanisms that characterize important medical conditions. New areas of expertise were developed, different specialties and subspecialties were established, and new important diagnostic tools were provided to patients. As a result, we have improved the treatment and cure of many different life-threatening diseases, but unfortunately, despite all this, morbidity and mortality are still very high worldwide for many important and different medical conditions [2].

The incidence and prevalence of cardiovascular diseases (CVDs), cerebrovascular pathologies, diabetes, cancer, and many other important health conditions are dramatically increasing worldwide and the health issues and the health economic impact of these diseases will increase in the future [3]. The progress achieved looks like a little stone thrown in the sea. Our goals seem at times impossible to achieve: after decades of effort, we are not able yet to prevent and control well many of the most common medical conditions impacting our societies.

We believe that at the present time we are at the beginning of a new era. If we invest well in our future with the correct tools, we will be able to foster significant improvement in medicine and science. Progress and technology should be combined with science and clinical practice, and the current vision and knowledge for each medical field should take into account the important new lessons we have learned.

At the beginning of our international adventure, we always focused on the most important unmet medical needs. We were developing scientific projects of the highest possible caliber, but this was not enough!

We soon realized that something was missing! We were focusing and sharing biased and incomplete information. We were not applying and taking into consideration the great importance of the new discipline of Gender-Specific Medicine (GSM) [4–6].

We have to admit our failures, we have to recognize our successes, and we have to immediately start to reconfigure our view of all the subspecialties of medicine by applying the advances in GSM.

As a nonprofit educational foundation actively contributing to support the knowledge of the scientific community, we believe that emphasizing GSM will reshape our ability to ensure better health for all societies. From our own experience, it was amazing when a few years ago, we had the opportunity to promote a specific scientific program dedicated to GSM. During the course of a beautiful sunny day in Matera, a typical southern city in Italy called the city of the Stones "la città dei Sassi," we were impressed by the importance of the topics presented and discussed in this important new medical field. The most well-known and recognized international opinion leaders shared their data. We were astonished to learn how gender-specific differences exist for: the prevalence, susceptibility, pathophysiology, and outcome of disease [6]. There is clear evidence for sex differences in CVD [7] and in its clinical manifestations and natural history [8,9].

In fact, erroneous assertions that heart diseases were a man's diseases emanated from research almost exclusively focused on male patients, with sex-specific data available only for few cardiovascular studies [10,11].

Furthermore, it is essential that research programs, models for drug development, and testing need to take into account gender- and sex-based differences [12]. Cardio-metabolic risk factors and diseases such as obesity, body-fat distribution, type 2 diabetes, hypertension, and sleep disorder have gender-specific differences [4–6]. Prediabetic women have greater endothelial dysfunction than men, as well as more hypertension and a greater degree of fibrinolysis/thrombosis [13]. Furthermore, impaired fasting glucose seems a stronger risk factor for recurrent cardiovascular events among women than in men [14]. Women have many sex-specific risk factors, which enhance CVD risk, even though they have less obstructive coronary artery disease (CAD), but more abnormal coronary reactivity that includes microvascular dysfunction with more symptoms, ischemia, and adverse outcomes [15].

We have understood the importance of GSM, we have found a new lens through which to read medicine. Why are men and women so different? Is a single different chromosome able to determine this? We are looking forward to hearing these important answers! Some are already there and available to everyone. It is important to leave the past behind and to enter the future taking into consideration this new important newborn discipline in every field of medicine [16,17]. It is a very simple concept, but due to the fact it is a young (although not teen anymore) established fundamental root, it is not yet fully translated into the clinical research, daily practice, and treatment of the different medical conditions.

We feel very lucky, because the Foundation had the opportunity to see immediately the crucial role of GSM in all the different areas of medicine. Therefore, the best real evolution and progress in medicine should be the recognition worldwide of this new discipline. Aging, diabetes, metabolic diseases such as osteopenia, obesity, body-fat distribution, cardiovascular and cerebrovascular disorders, cancer, hypertension, and sleep disorder—every single topic in medicine could be incredibly influenced by this new filed [18–20].

Any field should apply this new point of view, by critically looking backward to incomplete or inaccurate information and by using the new science to recalibrate our daily medical practice and knowledge. As a Foundation we are also working very hard supporting the inclusion of GSM in all our scientific programs and we are delighted to report a great deal of interest from all the different scientific medical communities. Through the Foundation we are committed to move forward in this important direction and we are open and ready to start any new collaboration and challenging project that could help the GSM to spread out around the world faster and deeper for an optimal contribution to medical knowledge. We strongly believe in the great importance of GSM and believe that it is essential to the improvement of medical care and the future of scientific investigation [21].

REFERENCES

[1] World Health Organization, editor. World health statistics 2016: monitoring health for the SDGs, sustainable development goals. Geneva: World Health Organization; 2016.

[2] Lozano R, Naghavi M, Foreman K, editors. Global and regional mortality from 235 causes of death for 20 age groups in 1990 and 2010: a systematic analysis for the Global Burden of Disease Study 2010. Lancet 2012;380:2095–128.

[3] Wang H, Naghavi M, Allen C, et al. Global, regional, and national life expectancy, all-cause mortality, and cause-specifi c mortality for 249 causes of death, 1980–2015: a systematic analysis for the Global Burden of Disease Study 2015. Lancet 2016;388:1459–544.

[4] Oertelt-Prigione S, Regitz-Zagrosek V, editors. Sex and gender aspects in clinical medicine. London: Springer; 2012.

[5] Regitz-Zagrosek Vera. Sex and gender differences in health. EMBO Reports 2012;13(7).

[6] Legato Marianne J. Principles of gender-specific medicine. 2nd ed. New York: Elsevier; 2010.

[7] Mozaffarian Dariush, Benjamin Emelia J, Go Alan S, et al. Heart disease and stroke-statistics. Circulation 2015;131:e29–e322.

[8] Regitz-Zagrosek V. Sex and gender differences in symptoms of myocardial ischaemia. Eur Heart J 2011;32:3064–6.

[9] Regitz-Zagrosek V, Oertelt-Prigione S, Seeland U, et al. Sex and gender differences in myocardial hypertrophy and heart failure. Circ J 2010;74:1265–73.

[10] Blauwet LA, Hayes SN, McManus D, et al. Mayo Clin Proc 2007 Feb;82(2):166–70.

[11] Melloni C, Berger JS, Wang TY, et al. Representation of women in randomized clinical trials of cardiovascular disease prevention. Circ Cardiovasc Qual Outcomes 2010 Mar;3(2):135–42.

[12] Regitz-Zagrosek V. Therapeutic implications of the gender-specific aspects of cardiovascular disease. Nat Rev Drug Discov 2006;5:425–38.

[13] Donahue RP, Rejman K, Rafalson LB, et al. Sex differences in endothelial function markers before conversion to pre-diabetes: does the clock start ticking earlier among women? The Western New York Study. Diabetes Care 2007;30:354–9.

[14] Donahue RP, Dorn JM, Stranges S, et al. Impaired fasting glucose and recurrent cardiovascular disease among survivors of a first acute myocardial infarction: evidence of a sex difference? The Western New York experience. Nutrit Metab Cardiovasc Diseases 2011;21:504–11.

[15] Shaw LJ, Bugiardini R, Merz CN. Women and ischemic heart disease: evolving knowledge. J Am Coll Cardiol 2009 Oct 20;54(17):1561–75.

[16] Legato MJ, Johnson PA, Manson JE. Consideration of sex differences in medicine to improve health care and patient outcomes. JAMA 2016;316(18):1865–6.

[17] Miller VM, Kararigas G, Seeland U, et al. Integrating topics of sex and gender into medical curricula—lessons from the international community. Biol Sex Differ 2016;7(Suppl 1):44.

[18] Bhupathiraju SN, Hu FB. Epidemiology of obesity and diabetes and their cardiovascular complications. Circ Res 2016 May 27;118(11):1723–35.

[19] Pucci G, Alcidi R, Tap L. Sex- and gender-related prevalence, cardiovascular risk and therapeutic approach in metabolic syndrome: a review of the literature. Pharmacol Res 2017 June;120:34–42.

[20] D Mozaffarian, EJ Benjamin, AS Go et al.,Heart disease and stroke statistics—2016 update: a report from the American Heart Association. Circulation 2016;133:e38–e360.

[21] Legato MJ. Correction: gender-specific medicine in the genomic era. Clin Sci Dec 15, 2015;130(2):125.

CHAPTER 10

Overcoming the Skepticism to Reach Gender Equity and Appropriateness in Pharmacological Response

Flavia Franconi[1,2]
[1]University of Sassari, Sassari, Italy; [2]National Institute of Biostructures and Biosystems, Sassari, Italy

ITALIAN GENERAL CONTEST

The Italian Constitution of 1948 affirms that women and men have the same rights. But it is only since the 1970's that women scored some major achievements with the approval in 1975 of the new family code. Indeed, after this, the legal and social status of Italian women changed rapidly [1]. It is also important to recall that Italian society is very much influenced by Roman Catholicism. Starting from St. Augustine and St. Thomas Aquinas, the Catholic Church accepts the view of Aristotle, which considers women intrinsically inferior to men. So it is not surprising that in 1930 Pope Pius XI condemned women's emancipation [2]. We had to wait until the encyclical *Pacem in Terris* to observe a different attitude in about the Catholic Church's attitude to women. This encyclical declares "Women are gaining an increasing awareness of their natural dignity. Far from being content with a purely passive role or allowing themselves to be regarded as a kind of instrument, they are demanding both in domestic and in public life the rights and duties which belong to them as human persons." [3].

Actually, particularly in the north of Italy [4], there is a growing acceptance of gender equity. However, in some parts of Italian society, gender roles are still traditional. This is particularly true in the south of Italy where women are still stereotyped as housewives and mothers. In Italy, as consequence, the number of unemployed women is 50% more than the European Union average employed [5]. However, occupational segregation, pay gaps, and glass ceilings are prominent issues in Italy as in many other OECD countries [6]. The Istituto Nazionale

The International Society for Gender Medicine. DOI: http://dx.doi.org/10.1016/B978-0-12-811850-4.00010-7

di Statistica e-book "*Come cambia la vita delle donne*" [5] shows that women's lives have changed greatly in recent years. For instance, women spend more money on culture versus men, and have a multitude of roles in the different phases of their life. However, the long economic crisis of Italy slows down women's progress. A consequence of the Italian economic crisis is the increase in the number of female workers who stop work after the birth of a child (women were more likely to do this than men even before the crisis). The crisis amplified the differences between the north and the south of the country.

It is also relevant and important to recall that there is a great deal of asymmetry of roles within the couple when both partners work full time. The couple with men described as "breadwinner" is more frequent than in other European countries [5]. Consequentially, women spend more time in the role of care giver. Because of this, there is an increasing delay in the timing of pregnancy and there are fewer babies born [5]. Signs of a positive change in income status are found among single women and women who lived as part of a couple [5]. Among the elderly, albeit with effort, a new woman emerges. In fact, highly qualified women break the stereotype of the older woman as poor, alone, in bad health: a burden for society [5].

MEDICAL CONTEST

The term "gender" comes from the Latin *genus* (descent, family, type). Recently, gender has been used to indicate social and cultural status, while the term "sex" indicates biological differences that involve hormones, genes, etc. The World Health Organization (WHO) declares that gender is "the socially constructed roles, behaviors, activities, and attributes that a given society considers appropriate for men and women" [7]. As is well known, prior to 2000 there was very little interest in gender medicine. Since 2000, however, WHO has included gender medicine in the Equity Act to state that equity should be applied to the access to care and should be promoted through the provision of appropriate medical care to all humans independent of sex. At the end of 1990s and in the beginning of the 2000s, the birth of the Partnership for Gender-Specific Medicine at Columbia University (1997), the Karolinska Institutet (2002), and the Charité Universitätsmedizin Berlin (2003) greatly increased the development of gender medicine in Italy and throughout the world. Beyond these centers, it is not possible to forget the role of the American Institute of Medicine of the National Academy of Sciences. In 2001, 2010, and 2012 the Institute of Medicine declared that being a woman or a man significantly influence the course of disease and should be considered in diagnosis and therapy [8–10]. Other important milestones for the incorporation of sex and gender aspects into preclinical and clinical research come from the following organizations: the Food and Drug Administration [11–14], the United State General Accounting Office [15,16], the International Conference on Harmonisation [17,18], the National Institute of Health [19,20], the Government of Canada [21], the Canadian Institute of Health [22], the Health Council for International Organizations of Medical Sciences [23], and the European Council Directive [24]. Through the work of these organizations, preclinical and clinical investigations began to systematically examine both the differences and similarities between women and men. It is necessary to emphasize that in medicine it is not easy to separate the influence of sex and gender on diseases. Social practices affect secondary sex characteristics, or the physiological and biological features that are commonly associated with maleness and femaleness. Vice versa, sex influences gender role [25]. In other words, sex and gender may work together. In 2003 the Italian Pharmacology Society created a working group in Gender Pharmacology pointing this out. Moreover, some attention was posed on women health by Italian Ministry of Health and Department of Equal Opportunities, and, in 1996, Dr. Modena organized "BenEssere Donna" in Modena; and in Napoli, Elvira Reale directed a center founded in 1987 on mental health focused on women. Finally, it is worth recalling, in recent times, the establishment of National Laboratory of Sex-Gender Medicine of the National Institute of Biostructures and Biosystems in Sardinia (2009).

BECAUSE I STUDIED GENDER PHARMACOLOGY

In this content, an area that merits increased attention is pharmacologic interventions. Drug treatment is the most common of therapies and its prevalence is still increasing. It is estimated that annual spending for drugs will reach $1.3 trillion in 2018 [26]. Heterogeneity in both efficacy and safety profiles of drug response is very high. In particular, it has been estimated that only 50%–75% of patients have benefits from the first drug offered in the treatment of diseases [27]. Actually, beyond drug-specific factors it is evident that pharmacological response depends on numerous factors (Table 10.1) and that gender and sex are two of them. Differences in the pharmacokinetics (absorption, metabolism, distribution, and elimination) in men and women were identified [28]. The pharmacodynamic differences are less well known but they are still emerging [29,30]. What is less known is the influence of gender on pharmacological response. Some studies [25,31] (and quoted literature) indicate that women are less likely than men to be adherent to chronic drug therapy, and they use more medications. Notably, it is proposed that the role of care giver could modify the adherence to a drug regimen [32]. Indeed, in women over 50 years of age, being a care giver seems to be a major barrier in the prevention for cardiovascular diseases [33]. Glaser et al. [34] show that the care giver role alters the response to pneumococcal vaccine. Interestingly, sex-gender of care providers may modify the adherence to pharmacological therapy [25]. Relevantly, in patients with heart failure, male physicians prescribe less medications and use lower doses in female patients in comparison to males. Female gender of physicians is an independent predictor of use of beta-blockers [35].

Table 10.1 Some Factors That May Affect Pharmacological Response Beyond Drug Factors

Physiological Factors	Genetic Factors	Environmental Factors	Cultural, Social Factors	Pathological Factors
Age	Receptors	Foods and beverages	Gender	Gastrointestinal diseases
Body dimension and composition and other differences	Enzymes that metabolize drug	Exposure to environmental chemicals	Stressors	Renal diseases
Ethnies	Deficiency of G6PD	Occupation	Social status	Diabetes mellitus
Sex		Drugs including hormones		Drug addiction
Timing				Allergy
Temperature				Heart failure
Pregnancy				
Puerperium				
Menstrual cycle				

Compared to the relatively biologically rich data-set that explains the drug activity, there is practically no information regarding how, and to what extent, cultural and societal settings affect pharmacological responses. However, there are data that suggest that they influence treatment outcome. For example, non-Caucasian subjects may be more responsive to placebo treatment than Caucasians [36]. Several studies also have elegantly demonstrated that the placebo effects, efficacy, and safety of drug are intimately influenced by the patients' cultural background.

In the mid-1990s I became completely aware that the health and medical data were based on the paradigm of "young, adult, male, white" men weighing 70 kg [25]. The West of Scotland Coronary Prevention Study (WOSCOPS) confirms this; it enrolled only men. The study shows that pravastatin in primary prevention reduces cardiovascular events in men within 6 months [37] and these results were translated to women without any direct study of their response to the drug. A confirmation of the androcentric view of medicine also comes from the Zolpidem story. Only recently, the Food Drug Administration recommended a different dose of Zolpidem, a hypnotic drug, for women and men [38], although the drug was on the market for several years. However, as early as 1932, Nicholas and Barron showed that the pharmacological action of barbiturates depends on sex: female rats required half the dose needed by male rats to induce sleep [39]. Inevitably, the simple extrapolation of data obtained in men to women and the ignorance of variables such as age, hormonal status, and culturally dictated differences between men and women leads to an inappropriateness of care and to worse safety profile in women compared to men [25,40].

Another important aspect of drug treatment is the patient's economic status. Poverty and low social status deleteriously affect health. They are associated with psychiatric and cardiovascular diseases, as well as diabetes mellitus [41–46]. For example, people with a lower social economic status smoke more than individuals with a higher social economic status [47], and it is well known that smoking is positively associated with mortality: it increases the risk of developing numerous diseases [48]. Indeed, smoking can interact with numerous drugs [49] and nicotine, the main psychotropic molecule present in the tobacco, is metabolized differently by the two sexes [50].

The above considerations support the conviction that gender research is a way to overcome the conventional reductionist approach based on one disease, one gene, one drug paradigm, attributing drug effectiveness to its action on a single target. Gender puts pharmacology into a multidimensional context, helping to contextualize the influence of environment throughout life including prenatal existence [51]. Thus, gender studies are a challenge because they require researchers coming from interdisciplinary and intersecting areas. Thus, gender studies are fabriques of expertise celebrating diversity and complexity. In doing that, gender studies change thinking, making it more precise and produce a more accurate understanding of the life experiences of women and men in society. The knowledge derived from gender studies may change society because gender medicine and gender pharmacology have a holistic approach. That holistic approach includes both men and women, and not a single target. It is evident that the female and male human body is greater than the sum of its parts. To increase the beneficial effect of clinical drug response, it is necessary to consider female and male biological organization in all its complexity and uniqueness. We must consider the stage of development, the impact of geography on phenotype, choose from an array of molecular targets including the microbiome and all omics. In other words, gender includes the biological networks and how they are perturbed by culture and society. Given the complexity resulting from summation of all potential biological interactions and social aspects, it is necessary to produce continual methodological improvements, which include a consider multidimensional approach to the human body—how it interacts with the environment, how it is modified by developmental stage and its modification by the drug. Although much work remains to do. I hope that gender pharmacology will rapidly develop to meet the challenges and arrive at precision medicine.

In conclusion, I am very glad to overcome the skepticism (when I started to study gender pharmacology, ironical smiles were very frequent) that I met when I started to be involved in gender studies.

REFERENCES

[1] European Parliament. <http://www.europarl.europa.eu/RegData/etudes/note/join/2014/493052/IPOL-FEMM_NT%282014%29493052_EN.pdf>; 2014.

[2] Pius XI. Casti Connubii. 1930.

[3] John XXIII. Pacem in Terris. 1963.

[4] Il mensile. <http://www.eilmensile.it/2012/02/20/sud-italia-questo-non-e-un-paese-per-donne/>; 2012.

[5] ISTAT. Come cambia la vita delle donne. 2015.

[6] OECD. <http://www.oecd.org/gender/closingthegap.htm>; 2012.

[7] World Health Organization, editors. Gender, women and health. WHO, Geneva; 2009.

[8] Wizemann T, Pardue M, editors. National Academy Press, Washington, DC; 2001.

[9] Wizemann T. Institute of Medicine; 2012.

[10] Institute of Medicine. Women's Health Research: progress, pitfalls, and promise; 2010.

[11] Food and Drug Administration, editor. Center for Drug Evaluation and Research Department of Health and Human Services; 1988.

[12] Food and Drug Administration. Guideline for the study and evaluation of gender differences in the clinical evaluation of drugs; 1993.

[13] Food and Drug Administration. FDA Action Plan to enhance the collection and availability of demographic subgroup data. <http://www.fda.gov/downloads/RegulatoryInformation/Legislation/SignificantAmendmentstotheFDCAct/FDASIA/UCM410474.pdf>; 2014.

[14] Food and Drug Administration. Safety and Innovation Act of 2012; 2012.

[15] General Accounting Office. FDA needs to ensure more study of gender differences in prescription drug testing. United States; 1992.

[16] General Accounting Office. Women's Health: NIH has increased efforts to include women in research. United States; 2000.

[17] International Conference on Harmonisation (ICH). Structure and content of clinical study reports (E3 Guidelines). <http://www.ich.org/fileadmin/Public_Web_Site/ICH_Products/Guidelines/Efficacy/E3/E3_Guideline.pdf>; 1995.

[18] International Conference on Harmonisation (ICH). Sex-related consideration in the conduct of clinical trials. <http://www.ich.org/fileadmin/Public_Web_Site/ICH_Products/Consideration_documents/ICH_Women_Revised_2009.pdf>; 2009.

[19] National Institutes of Health (NIH). Considering sex as a biological variable in NIH funded research. <http://grants.nih.gov/grants/guide/notice-files/NOT-OD-15-102.html>; 2015.

[20] National Institutes of Health (NIH); 1993.

[21] Canadian Governament. Guidance document: considerations for inclusion of women in clinical trials and analysis of sex differences. <https://www.canada.ca/en/health-canada/services/drugs-health-products/drug-products/applications-submissions/guidance-documents/clinical-trials/considerations-inclusion-women-clinical-trials-analysis-data-sex-differences.html>; 2013.

[22] Canadian Institute of Health. What a difference sex and gender make: a gender, sex and health research casebook; 2012.

[23] Council for International Organizations of Medical Sciences. International ethical guidelines for biomedical research involving human subjects. <http://www.cioms.ch/publications/layout_guide2002.pdf>; 2002.

[24] European Council. Directive 2001/20/EC of the European Paliament and of the Council on the approximation of the laws, regulations and administrative provisions of the Member States relating to the implementation of good clinical practice in the conduct of clinical trials on medicinal products for human use. <http://ec.europa.eu/health/files/eudralex/vol-1/dir_2001_20/dir_2001_20_en.pdf>; 2001.

[25] Franconi F, Campesi I. Sex and gender influences on pharmacological response: an overview. Expert Rev Clin Pharmacol 2014;7:469–85.

[26] The IMS Institute for Healthcare and Informatics. Global outlook for medicines through 2018. <http://www.imshealth.com/portal/site/imshealth/menuitem.762a961826aad98f53c753c71ad8c22a/?vgnextoid=266e05267aea9410VgnVCM10000076192ca2RCRD&vgnextchannel=736de5fda6370410VgnVCM10000076192ca2RCRD&vgnextfmt=default>; 2014.

[27] Spear BB, Heath-Chiozzi M, Huff J. Clinical application of pharmacogenetics. Trends Mol Med 2001;7:201–4.

[28] Franconi F, Brunelleschi S, Steardo L, Cuomo V. Gender differences in drug responses. Pharmacol Res 2007;55:81–95.

[29] Franconi F, Campesi I. Pharmacogenomics, pharmacokinetics and pharmacodynamics: interaction with biological differences between men and women. Br J Pharmacol 2014;171:580–94.

[30] Franconi F, Campesi I. Sex impact on biomarkers, pharmacokinetics and pharmacodynamics. Curr Med Chem 2016;23:1–15.

[31] Manteuffel M, et al. Influence of patient sex and gender on medication use, adherence, and prescribing alignment with guidelines. J Womens Health (Larchmt) 2014;23:112–19.

[32] Cherry N, Shalansky K. Efficacy of intradialytic parenteral nutrition in malnourished hemodialysis patients. Am J Health Syst Pharm 2002;59:1736–41.

[33] Mosca L, Mochari-Greenberger H, Dolor RJ, Newby LK, Robb KJ. Twelve-year follow-up of American women's awareness of cardiovascular disease risk and barriers to heart health. Circ Cardiovasc Qual Outcomes 2010;3:120–7.

[34] Glaser R, Sheridan J, Malarkey WB, MacCallum RC, Kiecolt-Glaser JK. Chronic stress modulates the immune response to a pneumococcal pneumonia vaccine. Psychosom Med 2000;62:804–7.

[35] Baumhakel M, Muller U, Bohm M. Influence of gender of physicians and patients on guideline-recommended treatment of chronic heart failure in a cross-sectional study. Eur J Heart Fail 2009;11:299–303.

[36] *Psychopharmacology and psychobiologyof ethnicity.* Edited by K.M. Lin, R.E. Poland & G. Nakasaki, 1993, American Psychiatric Press, Washington, DC.

[37] Shepherd J, et al. Prevention of coronary heart disease with pravastatin in men with hypercholesterolemia. West of Scotland Coronary Prevention Study Group. N Engl J Med 1995;333:1301–7.

[38] Kuehn BM. FDA warning: driving may be impaired the morning following sleeping pill use. JAMA 2013;309:645–6.

[39] Nicholas JS, Barron DH. The use of sodium amytal in the production of anesthesia in the rat. J Pharmacol Exp Ther 1932;46:125–9.

[40] Franconi F, Campesi I, Occhioni S, Antonini P, Murphy MF. Sex and gender in adverse drug events, addiction, and placebo. Handb Exp Pharmacol 2012;107–26.

[41] Krantz DS, et al. Extent of coronary atherosclerosis, type A behavior, and cardiovascular response to social interaction. Psychophysiology 1981;18:654–64.

[42] Muller JE, Abela GS, Nesto RW, Tofler GH. Triggers, acute risk factors and vulnerable plaques: the lexicon of a new frontier. J Am Coll Cardiol 1994;23:809–13.

[43] Carney RM, et al. Depression, heart rate variability, and acute myocardial infarction. Circulation 2001;104:2024–8.

[44] Elovainio M, et al. Socioeconomic differences in cardiometabolic factors: social causation or health-related selection? Evidence from the Whitehall II Cohort Study, 1991–2004. Am J Epidemiol 2011;174:779–89.

[45] Ghiadoni L, et al. Mental stress induces transient endothelial dysfunction in humans. Circulation 2000;102:2473–8.

[46] Veronesi G, et al. Gender differences in the association between education and the incidence of cardiovascular events in Northern Italy. Eur J Public Health 2010;21:762–7.

[47] Giskes K, et al. Trends in smoking behaviour between 1985 and 2000 in nine European countries by education. J Epidemiol Community Health 2005;59:395–401.

[48] Kulik MC, et al. Educational inequalities in three smoking-related causes of death in 18 European populations. Nicotine Tob Res 2014;16:507–18.

[49] Kroon LA. Drug interactions with smoking. Am J Health Syst Pharm 2007;64:1917–21.

[50] Agabio R, Campesi I, Pisanu C, Gessa GL, Franconi F. Sex differences in substance use disorders: focus on side effects. Addict Biol 2016;21:1030–42.

[51] Franconi F, Rosano G, Campesi I. Need for gender-specific pre-analytical testing: the dark side of the moon in laboratory testing. Int J Cardiol 2015;179:514–35.

CHAPTER 11

Gender Medicine in Italy: The Point of View of Maria Grazia Modena

The Story, Limits, and Plans: My Personal Story

Maria G. Modena and Valentina Martinotti
University of Modena and Reggio Emilia, Modena MO, Italy

THE STORY

I do not think it would be presumptuous to say that in a sense gender medicine (GM) in Italy was born in association with the group of which I have been part of since the 90s. My interest in GM began in fact in the 90'sas a woman doctor taking care of women with heart disease. In 1996 I founded the first women's clinic in Italy (BenEssere Donna) dedicated, at the beginning, to menopause-related diseases, reflecting my clinical research, born with gynecologists in the era of hormonal replacement therapy (HRT) which was very popular at that time in Italy and Europe. HRT was considered a prevention tool, much more popular and "scientifically" relevant than the issue of heart disease in women; this because of market's interest in hormones. Starting from this and publishing some papers on HRT and risk factors (RFs), I focussed my interest on cardiovascular disease in women.

In 1999 I was invited to join the association "Una Salute a Misura di Donna" (Tailored Health for Women) created by Elvira Reale (psychologist chair), Patrizia Romito (psychologist), Nadia Pallotta (gastroenterologist), Giuseppina Boidi (psychiatrist), Silvana Salerno (epidemiologist), Irene Figà Talamanca (occupational health), Adriana Ceci (pediatrician), Laura Corradi (sociologist), and myself. It was fascinating, and in a sense "romantic," the beginning of our association, since Doctor Reale invited us to a convent of nuns on the hills of Naples for 3 days of full immersion in GM. The meeting was a success and very motivating. Our adventure, therefore, began with few funds

The International Society for Gender Medicine. DOI: http://dx.doi.org/10.1016/B978-0-12-811850-4.00011-9

and great enthusiasm and we decided to develop research in RFs related to a woman's daily life.

The *general aim* of our group was to create a unified field of observations about the most common diseases affecting women, to point out research bias against women and to propose guidelines of intervention for a women-oriented healthcare system.

The *specific aims* were

- to present women's health problems with an integrated approach taking into account, but not exclusively emphasizing, women's biological cycles;
- to discover inequalities between men's and women's treatments concerning relevant diseases in the population, particularly mental, cardiovascular, oncologic, gastroenterological, and stress/violence-related diseases;
- to point out the deficiency of the medical organization of services and planning to take in consideration all women's health needs, and not just the ones related to their reproductive function.

The group, after 3 years of work, in 2001 published its first results in a "Report on Women's Health" (along with the cooperation of Italian Ministry of Equal Opportunities), and later in 2003, realized a "Guide for Women and Professional" (along with the cooperation of the Italian Ministry of Health) [1,2]. These reports highlighted a number of relevant problems and prejudices which limited research and clinical work in behalf of women's health. I thought it was important to include part of those documents in this chapter, because they are still functional and are milestones in the history of GM in Italy; all references mentioned are included in the two reports. An update of our report is in progress and will be published soon. We described at that time the incidence of relevant new RFs, such as environment, violence, work and gender difference in cardiovascular diseases, mental illness, malignancies, AIDS- and stress-related diseases. The lack of gender-specific health data is the first problem to be mentioned. It is well known that women have longer life expectancy than men in similar socio-economic positions. Nevertheless, they suffer from poorer overall health because of age, biological and specific RFs, and loneliness. According to the data of the National Statistical Institute, life expectancy in Italy is 82 years for women and 75.8 for men. This gap between women and men decreases if we consider life expectancy after the age of 65 and life expectancy free from disability. At age 65, life expectancy is 20 years for women and 16 for men, which is a gap of 4 years, but if we consider life expectancy free from disability, this gap is reduced to 1 year only. At age 75 the difference in life expectancy between women and men free from disability is irrelevant; only 0.3 years (according to Ministry of Health's data). Moreover, compared to men, women report more comorbidities (particularly arthritis, osteoporosis, hypertension, and depression); survey data on self-reported state of health show that a wider share of women than men reports at least one chronic disease, and in 23 out of 28 categories of diseases, the prevalence is greater among women (National Statistical Institute). The report provided data on male and female prevalence and incidence rates for several diseases. Where national official data were not available, the experts of the group have drawn from international data and from data originating from their own research and clinical work. More men die from lung cancer but in Italy recently women's mortality rate for this cause increased by 18%, while the men's rate decreased by 4%. Moreover, a lung cancer epidemic for women is foreseen for the coming decade. Mortality among women for AIDS has reached the male rate; indeed, AIDS constitutes the main cause for death among women in the age bracket 15–44, and in this age bracket their mortality rate is higher than men's rate. As the World Health Report shows, all types of mental disorders, except for alcohol and drug abuse, are increasing and are more common among women; in particular, unipolar depression ranks fourth for women and eight for men in terms of the main causes of disease burden throughout the world. Depression is 2–3 times more common among women compared to men and is the main cause for disability among women in the age bracket 15–44. Despite the fact that depression is more common among women, there are no gender-oriented research and intervention programs for its prevention and treatment in our country. Until recently schizophrenia was more commonly diagnosed in men, but recent data show a slightly higher incidence among women, who have less socially undesirable traits than men. Today also the rate of death from cardiovascular diseases is higher for women than for men. In Italy, according to the data of the National Statistical Institute, in 2000 (confirmed in 2008 and in 2016) cardiovascular diseases were the cause of death in 48% of adult women and in 39% of adult men. Hypertension, the most important RF, is more frequent in males up to thc age of 45, while in females it is more frequent from the age of 50. A recent report on women of the age bracket 45–60 showed that 38.2% suffer hypertension; however, it is striking that only 34% receive adequate treatment.

In spite of the fact that there are striking differences by gender for important diseases, in Italy a structured national system capable of monitoring social differences in health is not available; more specifically, except for cancer incidence and mortality, data and research on gender difference are lacking and health data are not systematically collected and disaggregated by sex. The information provided by hospitals and local health units is incomplete, not divided by sex and often unreliable. There are no clear national norms for including gender as a basic variable in all data collection.

Another important problem underlined in the reports is the lack of gender sensitive research on occupational and environmental RFs. Little attention has been given to RFs for depression, coronary artery disease, breast and uterine cancer since these diseases have been up to now considered to have mainly hormonal etiology. Let us make some examples. Tobacco smoking is the main RF for lung cancer. Women start smoking earlier than men and have more difficulty quitting. However, primary prevention of tobacco smoking is gender blind, i.e., aside from the period of pregnancy, specific messages aimed at girls and women are lacking. Instead an effective primary prevention should investigate gender and socio-cultural factors associated with smoking initiation and cessation.

In the field of mental health, research on RFs is mainly oriented toward the evaluation of biological-hormonal factors, generally omitting for females (but not for males) the investigation of psycho-social and work factors. There seems to be a gender bias in psychiatry since it often cites hormonal variations as the major RF for depression and other psychiatric disorders in women. As in general medicine where there has been an improper process of medicalization of the physiological stages of a woman's life, there has also been in psychiatry a process of psychiatrization of female physiology, particularly true for depression. Among women there has been a clear under-evaluation of environmental and psycho-social factors as well as the impact of daily life on their mental health. Similarly the research on RFs for coronary artery diseases, in women there is an over-evaluation of biological and hormonal factors and an under-evaluation of environmental and stress factors; stress is considered the main RF for ischemic heart disease only for men.

Occupational health remains an area where we lack knowledge about occupational hazards and their effects on women's health; often this type of data is not disaggregated by sex or there is insufficient details on women. However, in many occupations that may be considered "female" hazards are very high, in the health sector 54% of accidents involve women. As more women work, occupational injuries are increasing among them (+8.4% from 2000 to 2003) while decreasing among men (−9.8% during the same time period). More studies are needed to investigate whether occupational exposures have different effects on female workers than on male workers. One example is heavy physical work, which is less well tolerated by women. This is partly due to the fact that work organization is based on standard measures designed for male workers. Italian women suffer more frequently musculoskeletal disorders, compared to men, even in the same work environment. This is because women are given repetitive tasks in fixed and inadequate ergonomic positions which put excessive pressure on their smaller and more vulnerable muscles (Talamanca, 2001, 2003, confirmed in 2008). In studying RFs for cancer the type of work is always considered for men while the same emphasis has not always been taken into account for women. However, recent studies have shown the correlation between some types of tumor (kidney, lung, leukemia, lymphoma) and exposure to solvents and hydrocarbon among female workers. As we know a very important RF for women is the stress and strain linked to the multiple work load for the family and for the labor market. Our group underlined that in medical research and clinical practice, great importance is given to work as a major RF in the analysis, prevention, and treatment of diseases in men, but for women little attention is given to this factor or to other life conditions. In particular, little or no attention is given to the coexistence for women of a plurality of roles, responsibilities, and tasks linked to their professional and family life. The risk of physical and psychological burn-out is neglected, and parameters capable of measuring the hazards and satisfaction of family work are not available. This "invisible work," indeed, goes far beyond what is commonly called housework, but involves also the production of goods and services for family consumption, the bureaucratic activities necessary to use the public services, as well as the activities necessary for the education, socialization, and caring of the children, the elderly, the ill, and disabled members of the family. In this respect—due to the lack of public services and to the traditional division of roles within the family—Italian women's work burden seems to be particularly high. The study of family use of daily time has underlined that on the average; Italian women spend 5 h a day in paid work and 8 h in nonpaid work, while men spend 8 h in paid work and 1 h in nonpaid work. Moreover, men's pattern in the use of time does not vary significantly by the different type of family they live in, while women's use of time is significantly affected by the presence of children in the family and of a partner as well. In this respect, women with children seem to be advantaged by the absence, rather than by the presence, of a partner in the family. In the end, as the chairwoman of the Italian Equal Opportunity Commission has underlined, considering both paid and nonpaid work, more than 50% of Italian women work 60 h a week and over one-third work 70 h a week, while less than one-third of men work more than 60 h a week (Talamanca, 2003, confirmed in 2008 and in 2016). Thus family work and its character of caring for others (as opposed to caring for oneself) can well be considered as a major RF for women's health. As a matter of fact, research on stress and women's daily life suggests a relation between hypertension, breast carcinoma, and depression and the increase in family work responsibilities. Violence is a frequent experience in the life of women and is considered a RF for poor health among women. According to an Italian study on a sample of users of various socio-health services, 1 out of 10 women in the sample and 18% of those in the age bracket 18–24 had experienced physical or sexual violence in the 12 months preceding the survey. The perpetrators of the violence are almost always men close to the women: her partner, former partner, father, brother, school mates, or work colleagues.

Finally a very important problem that was addressed by the group was the exclusion or insufficient presence of women in clinical trials conducted to test new pharmaceutical products, where mainly adult males are enlisted. Moreover, even when women are employed in adequate numbers in these studies, the data are not as a rule analyzed by sex.

BARRIERS

Unfortunately, our work, initially very prolific, had little success recently, because both the Ministers of Equal Opportunities and Health were changed and both successors were definitely not interested in the problem. This is a typical Italian problem: politics has too much influence on social orientation and health care (and governments in Italy have changed and change too frequently!!!). At the beginning of 2000, each of us started personal research on gender aspects in her own field of interest, sharing ideas and experiences in national meetings every year. Here we soon met another barrier since being the first to deal with GM in Italy, and all women, we were considered (or rather confused) with feminists. Our congresses were, therefore, attended by few doctors (and those who attended were almost women) for several years and never crowded. The third barrier was our attempt to act on the Italian Pharma Agency (AIFA) to promote greater enrollment of women in clinical trials, following the Food and Drug Administration's lead and, then, the European Medical Agency's recommendations that, to endorse a clinical trial, the trial should contain at least 25% of women. AIFA welcomed and accepted our request, reporting from 2001 to 2014 a progressive increase in specific women's studies, more focused on Phase III (enrollment), absence of Phases I and II, and little relevance for Phase IV. But analyzing the trials we observed that most of the new studies including women concerned drugs used for female illnesses such as: breast cancer, osteoporosis, genitourinary, and fertility (ovarian stimulation and assisted fertilization) drugs.

THE LAST, PERSONAL, BARRIER: MY STORY AS AN EXAMPLE OF "GENDER RELATED" STORM

My clinical research started, as I said, from investigating HRT (transdermal estradiol, in selected women with menopausal symptoms, improves endothelial dysfunction and metabolic profile, through an improvement of "flow-mediated vasodilation") and moved to the study of endothelial dysfunction [3–5]. Menopause per se, I reported, is not a disease, but always coincides with endothelial dysfunction. RFs may induce atherosclerosis, which produces endothelial dysfunction, independent of menopause, but, together, they accelerate the process

[6–13]. Analyzing data from the Womens' Clinic my interest focused on hypertension in all phases of woman's life: physiopathology, diagnosis, and treatment. Endothelial dysfunction, in different degrees, represents the basis of hypertension during pregnancy, in preeclampsia/eclampsia and in postmenopausal life with or without the metabolic syndrome. The "Women's Clinic" over time changed and became the reference point for all aged women with RFs or cardiovascular diseases and data were published on specific results [14–26]. In the meantime, my career and visibility also grew. I became the Head of Cardiovascular Diseases Unit and Chief of Department of Emergency-Acceptance, then the first woman President of the Italian Society of Cardiology. My interest in GM continued, but I started to monitor women's careers in medicine, realizing how many obstacles I had and I was meeting as a leading woman in cardiology. This is well summarized in the interview by Barry Shurlock on Circulation (European Perspectives in Cardiology) [27] "Viewpoint of Maria Grazia Modena: politics is never far from life in Italy." This observation applies as much to medicine as it does to great national events "- Dr Maria Grazia Modena, a professor at the University of Modena and Reggio Emilia, looks back on her 2-year term (from 2004 to 2006) as the first female President of the Italian Society of Cardiology (ISC), when she carried out what she frankly calls a revolution in its constitution. 'As soon as I became president of the society, I decided to review the rules and how it was run. I soon realised there had to be changes, with the agreement of the board, of course. There were members who had held the same position for 12 years or more'. She explains, 'An effective society has to have a rapid turnover of officials—the old have to give way to the young. There were 50 people running the society, sitting every month around a table—it was too many, so we reduced it to just 11 -and consequently decreased the expenses of the society, also stopping the custom of supplying ISC members with luxurious hotel accommodations for boards members staying during our monthly meeting'. She was also keen to motivate more young cardiologists, so in addition to the annual meeting, she instituted 2 new types of meeting, held twice a year. Still to be achieved, but close to her heart, is another revolution: to gradually join together the Italian Society of Cardiology (Academic Cardiologists) and its sister society the National Association of Hospital Cardiologists. Dr Modena would like to see both organisations actively participating in the Italian Federation of Cardiology. She believes that it does not serve the cause of Italian cardiology well to have 2 societies doing very similar jobs that would be better performed if they combined forces. Asked to define her main research interest, Dr Modena says, 'It is heart disease and women. By this I mean not only the better care of women with heart disease, but also the better representation of women as practising cardiologists'. In 1999, she published a survey of women and cardiology in Italy, and later she became chair of the Women in Cardiology Committee (WIC) of the European Society of Cardiology. In her final report, Dr Modena outlined the status of women within the European Society of Cardiology, pointing out that although 26% of members overall were women (17% in the Mediterranean region and 40% in the former Eastern bloc), they were underrepresented within the Society. Only 8 of 31 European Society of Cardiology committees had any female members, and only 10% of working groups included women...."

WIC was suppressed since the new ESC's President at that time decided the committee was useless and, on the other hand, the president of ISC who succeeded me in 2007 reintroduced all I had changed and reformed by declaring that I was a communist!!!!!

Coming back to my work in Modena, I was appointed to a board that had to face healthcare problems for all the hospitals in the city and around. I was the only woman and I was committed to make proposals to reduce healthcare costs. I proposed to unify the Cardiology Units of the four hospitals, too many for our city, into one large department, with specific diagnostic/therapeutic pathways for myocardial infarction, heart failure, heart surgery, Women's Clinic, etc. The others suggested nothing! I was accused of attempting to control everything and my proposal in 2011 was rejected. Actually, because of the severe economic crisis, my project is now in progress!!!

In June 2012 a major earthquake hit my region and destroyed my home birthplace and in November of the same year, when my career was at its peak, I was involved in a major legal scandal (linked to political struggles and local envy) with accusations of criminal association, abuse of power, corruption, and fraud!!! A bad gender-related story, because I am sure it would never have happened to a man. In December 2016, after 4 years, I have been absolved of everything with the explanation that I had no motivation for having committed any

Figure 11.1 Tel Aviv, Israel, September 2014, 7th International Congress IGM. From left to right: myself, Professor Marek Glezerman, Professor Marianne Legato and Professor Jeanette Strametz Juranek.

crime. During those years to be part of the board of International Society for GM and have the esteem and the respect of all members and mainly of Professor Marianne Legato, our illustrious Editor, and of Professor Marek Glezerman, our President (Fig. 11.1), was vital for me.

In the meantime I have continued my clinical research on GM focusing on gender difference in the effects of posttraumatic stress for people hit by the earthquake [28] and I wrote two books on my story [29,30].

A geological earthquake and a psychological earthquake have profoundly changed my life!!!

LIMITS AND PLANS FOR THE FUTURE

GM has many barriers in Italy, some of which have already been addressed by our Association. In 2008 the Ministry of Health asked and published an update [31] of our previous report and, surprisingly, most of our data on gender difference significantly overlapped with the previous report as well as our recent research; 2016 data, do not deviate much from the past. This sounds to me as the demonstration of a lack of progress. But the most relevant limit to the expansion of

GM in Italy is excessive fragmentation. In fact, in the last 10 years, many associations/working groups have emerged, organizing meetings but not doing research. Finally a gender-based law is still awaiting government approval for years.

Therefore the most urgent plan is, in my opinion, (1) try to join forces; (2) to introduce GM in the core curriculum of medical school to raise awareness of the gender concept.

The risk otherwise, at least in Italy, is to see GM relegated to a niche topic for a few elected women—a luxury niche, just like a boutique.

REFERENCES

[1] Aa Vv. Atti del gruppo di lavoro "Medicina Donne Salute", Ministero Pari Opportunità. Una salute a misura di donna. Roma: E. Reale; 2001.

[2] La Salute Mentale delle Donne, Roma; 2003. Online report available from <http://www.palazzochigi.it/cmparita/commissione/attivita/pubblicazioni/mente_cuore_braccia/>.

[3] Rossi R, Origliani G, Modena MG. Transdermal 17-β-estradiol and risk of developing type 2 diabetes in a population of healthy, nonobese postmenopausal women. Diabetes Care 2004;27(3):1–5.

[4] Rossi R, Bursi F, Veronesi B, Cagnacci A, Modena MG. Effects of progestins on estrogen-induced increase in C-reactive protein in postmenopausal women. Maturitas 2004;4:315–20.

[5] Modena MG, Sismondi P, Mueck AO, Kuttenn F, de Lignieres B, Verhaeghe J, et al. New evidence regarding hormone replacement therapies is urgently required. Transdermal postmenopausal hormone therapy differs from oral hormone therapy in risks and benefits. The TREAT Collaborative Study Group. Maturitas 2005;52:1–10.

[6] Rossi R, Chiurlia E, Nuzzo A, Cioni E, Origliani G, Modena MG. Flow-mediated vasodilation and the risk of developing hypertension in postmenopausal women. J Am Coll Cardiol 2004;44(8):1636–40.

[7] Rossi R, Cioni E, Nuzzo A, Origliani G, Modena MG. Endothelial-dependent vasodilation and incidence of type 2 diabetes in a population of healthy postmenopausal women. Diabetes Care 2005;28:702–7.

[8] Ratti C, Chiurlia E, Grimaldi T, Barbieri A, Romagnoli R, Modena MG. Breast arterial calcifications and coronary calcifications: a common link with atherosclerotic subclinical disease? Ital Heart J Suppl 2005;6(9):569–74. Review. Italian.

[9] Zanchetti A, Facchetti R, Cesana GC, Modena MG, Pirelli A, Sega R. Menopause-related blood pressure increase and its relationship to age and body mass index: the SIMONA epidemiological study. J Hypertens 2005;23(12):2269–76.

[10] Modena MG. Acute myocardial infarction—are women different? Int J Clin Pract 2005;59(1):2–5.

[11] Rossi R, Turco V, Origliani G, Modena MG. Type 2 diabetes mellitus is a risk factor for the development of hypertension in postmenopausal women. J Hypertens 2006;24(10):2017–22.

[12] Nuzzo A, Rossi R, Modena MG. Endothelial dysfunction in postmenopausal women and hypertension. Editorial. Women's Health 2007;3(5):13–21.

[13] Rossi R, Nuzzo A, Origliani G, Modena MG. Prognostic role of flow-mediated dilation and cardiac risk factors in post-menopausal women. J Am Coll Cardiol 2008;51:997–1002.

[14] Rossi R, Nuzzo A, Origliani G, Modena MG. Metabolic syndrome affects cardiovascular risk profile and response to treatment in hypertensive postmenopausal women. Hypertension 2008;52(5):865–72.

[15] Rollini F, Mfeukeu L, Modena MG. Assessing coronary heart disease in women. Maturitas 2009;62(3):243–7.

[16] Leuzzi C, Sangiorgi GM, Modena MG. Gender specific aspects in the clinical presentation of cardiovascular disease. Fundam Clin Pharmacol 2010;24:711–17.

[17] Nuzzo A, Rossi R, Modena MG. Hypertension alone or related to the metabolic syndrome in postmenopausal women. Expert Rev Cardiovasc Ther 2010;8(11):1541–8.

[18] Leuzzi C, Marzullo R, Tarabini Castellani E, Modena MG. Cardiovascolar risk factors. InTech; 2011. p. 327–36. 14, Gender-Specific Aspects in the Clinical Presentation of Cardiovascular Disease Minerva, Milano

[19] Nuzzo AC, Modena MG. Gender differences in cardiovascular risk and metabolic syndrome. J Intern Emerg Med Suppl XI, 2012;22–6.

[20] Volpe R, Sotis G, Gavita R, Urbinati S, Valle S, Modena MG. Healthy diet to prevent cardiovascular diseases and osteoporosis: the experience of the 'prosa' project. High Blood Press Cardiovasc Prev 2012;19(2):65–71.

[21] Maurice MC, Mikhail GW, Mauri i Ferré F, Modena MG, Strasser RH, Grinfeld L, et al. SPIRIT women, evaluation of the safety and efficacy of the XIENCE V everolimus-eluting stent system in female patients: referral time for coronary intervention and 2-year clinical outcomes. EuroIntervention 2012;8:325–35.

[22] Sangiorgi G, Roversi S, Biondi Zoccai G, Modena MG, Servadei F, Ippoliti A, et al. Sex-related differences in carotid plaque features and inflammation. J Vasc Surg 2013;57(2):338–44.

[23] Giubertoni E, Bertelli L, Bartolacelli Y, Origliani G, Modena MG. Parity as predictor of early hypertension during menopausal transition. J Hypertens 2013;31:501–7.

[24] Barbieri A, Mantovani F, Bursi F, Ruggerini S, Lugli R, Abdelmoneim SS, et al. Prognostic value of a negative peak supine bicycle stress echocardiography with or without concomitant ischemic stress electrocardiographic changes: a cohort study. Eur J Preven Cardiol 2015;22(5):636–44.

[25] Modena MG. Hypertension in postmenopausal women: how to approach hypertension in menopause. High Blood Press Cardiovasc Prev 2014;21(3):201–4.

[26] Modena MG. Aging. In: Preedy V, editor. Oxidative stress and antioxidants. 1st ed. Amsterdam: Elsevier; 2014: 22 Hypertension, menopause and natural antioxidants in foods and diet.

[27] Viewpoint: Maria Grazia Modena MD, FESC. Interview by Barry Shurlock. Abstract. Circulation european perspectives; 2007 October 2;11 6(14):f79–81.

[28] Modena MG, Pettorelli D, Lauria G, Giubertoni E, Mauro E, Martinotti V. Gender differences in post traumatic stress: From pathophysiology to the case of a natural catastrophic event. Available from: BioResearch Open Access; 2017, 6.1 <http://online.liebertpub.com/doi/10.1089/biores.2017.0004>.

[29] Modena MG. The cardiology case. My life. My truth. E-book, Edizioni il Fiorino; 2013.

[30] Modena MG. Il Caso cardiologia. La Verità. Modena, Edizioni il Fiorino; 2017.

[31] Modena MG, Reale E. Lo Stato di Salute delle Donne in Italia: primo rapporto sui lavori della Commissione "Salute delle Donne"; 2008, p. 1–205.

CHAPTER 12

Center for Gender Medicine Karolinska Institutet (KI) Since 2001

Karin Schenck-Gustafsson
Center for Gender Medicine, Department of Medicine, Karolinska Institutet and Karolinska University Hospital, Solna, Stockholm, Sweden

HOW IT STARTED

As early as the 1980s it was very clear to me that sex differences existed in cardiovascular disease (CVD). For many years I had worked full time as a clinical cardiologist at Huddinge University Hospital south of Stockholm. Then in 1986 I moved over to the Thoracic Clinics at the Karolinska University Hospital in Stockholm (These two hospitals have merged to become one of the largest in hospitals in Europe, called NKS (New Karolinska Hospital)). I was appointed head of the coronary care unit (CCU) and when meeting the patients there I began to consider the textbook statement that myocardial infarction (MI) was a man's disease: half of my MI patients were women! I also noted that women patients with unstable angina pectoris or small infarctions who had "normal" angiography were treated as "simulants." We even had a French radiologist calling them "des simulants coronaires." Sometimes they were even referred to psychiatry. When discharged from the ward out into the "black hole" all their heart medications were stopped. As a consequence their symptoms worsened. Now, 30 years later, we know that these women suffered from microvascular dysfunction or coronary artery spasm.

At that time there was a huge underrepresentation of women in cardiovascular trials in spite of the fact that CVD was the main cause of death for women even then. Nowadays, in Europe including Sweden, more women than men die of MI but women are still underrepresented in cardiovascular research. Furthermore, when research results are presented, they were and are still seldom analyzed from a sex and gender perspective. Women are about 10 years older than men when suffering from MI. Up

The International Society for Gender Medicine. DOI: http://dx.doi.org/10.1016/B978-0-12-811850-4.00012-0

to half of women with MI did not have the typical MI symptoms and 97% of animal experiments studying pain were performed in male animals and pain mechanisms were found to differ between the sexes.

It was also shown that woman more often than men had another type of heart failure with so-called preserved left ventricular function but most trials studied and still study the male type of heart failure with left ventricular systolic dysfunction. Another example was that women much more frequently than men suffer from a lethal rhythm disturbance called "torsade de pointe-ventricular tachycardia" induced by different medications. Curious enough, this was the subject of my PhD thesis in 1982, but at that time I was not aware of the sex differences.

So, to investigate this, to increase the knowledge about sex and gender differences in order to create better diagnostics and treatment for both men and women, I and my co-workers started the NIH funded Stockholm Female Coronary Artery Study (KOK) in 1991. We aimed to describe unique female cardiovascular risk factors and explore pathophysiology and optimal diagnostic procedures for women. We found that in women with acute coronary syndrome in Stockholm more than 25% had "normal" angiography without obstructive coronary artery disease. We also performed quantitative coronary angiography and intravascular ultrasound and could visualize early atheromatosis in normal looking coronary vessels [1]. Psychosocial risk factors, especially marital stress, emerged as especially important for women [2–25]. This was the start of my interest in sex differences in the area of ischemic heart disease and it was followed by exploring reproductive factors and their connection with cardiovascular disease.

I was also interested in the unique female features of cardiac rhythm disturbances which was the topic of my dissertation.

After having achieved funding from Karolinska University Hospital in 1998, I started preparing a facility for women's health. Then I suffered 6 months of significant illness. Instead of being discouraged, I was empowered and found out that I had to change strategy. So, in the year 2000 I did a tour of the United States to visit Centers of Women's Health of Excellence and visited New York, Los Angeles, Winston Salem, and New Orleans. In New York I visited Prof. Marianne Legato's office at Columbia University for her Partnership for Women's Health and became very inspired.

I realized that this area of medicine was not only about women's health. Men's health needed to be researched from a sex and gender perspective. In psychiatry a debate was going on about depression in men arguing that diagnostic tools might not be appropriate for men leading to underuse of antidepressive medication and as a result much more completed suicides among male patients than female. I also realized that the problems I was seeing in cardiology and psychiatry also existed in many other areas. There was a lack of sex and gender awareness in both clinical and preclinical medicine in general. I considered this to be dangerous since it can lead to wrong or less than optimal treatment for both genders and as a consequence, increased patient morbidity and mortality. Another aspect was that when comparing how a certain disease might be experienced differently among men and women, it can lead to deeper understanding of the cause of the disease.

My next step was to knock on the door of the Dean of Research at Karolinska Institutet (KI), Prof. Harriet Wallberg-Henriksson (later the president of KI). She found my ideas about a Center for Gender Medicine very interesting. We started fundraising and managed to get funded by 2.5 million Euros, to start up. The money came from the Wallenberg Foundation, AFA insurance company, and KI. The decision to start a center was supported by the Board of Research at KI in 2001 and a Director was to be appointed. By then Prof. Wallberg-Henriksson had moved to the Swedish Research Council and a new Dean of Research was in charge. He wanted to appoint a candidate informally (who happened to be a friend of his) as leader of the center. I was very disappointed but managed to get the position advertised and among 8 applicants I got the position. April 2002 marked the start of my directorship of the. Center for Gender Medicine (CfG) This new venture was among the first in the world and it became one of several KI research centers. It is structured according to the existing KI regulations of research centers. A board was appointed with representatives from other boards like the Boards of Research and Education together with laymen. Zsuzsanna Wiesenfeld-Hallin, Professor of neurophysiology, was appointed as the first chair. CfG has mainly been supported financially each year by KI, Stockholm County Council with additional support by private funding and NGO's like the 1.6 million Club for Women's Health.

As a token of success I have been awarded the H.M. the King's Medal for starting gender medicine in Sweden, Florio Maseri Award in Gender Cardiology by American College of Cardiology, and also an Award from the Swedish Association of Medicine in addition to many other awards. I received an honorary doctorate from the Medical University in Innsbruck, Austria for being the first who opened a Center for Gender Medicine in Europe.

CREATING A CENTER FOR RESEARCH AND EDUCATION

The Center had an inauguration party at the Nobel Forum at KI with HM, Queen Silvia, as guest of honor and the Queen was also present at the 10th anniversary in 2011. She has been a keen supporter during the years. Since the start, the Center drew a great deal of attention from Swedish and European media. The establishment of the center also initiated a debate in Sweden about sex and gender. We were criticized by people from nonmedical faculties who thought that we paid too much attention to biology and by researchers who wanted to attenuate the role of biology and focus more on gender theories. Our conviction is that there are both sex and gender dimensions in health and disease and that there are no strict borders between the two. Biology is an integral part of society.

During the first years we decided to stimulate research in gender medicine at KI and we funded the best applications for research support. After independent referee's evaluation 25,000–50,000 Euros per research group and PhD projects were distributed. That resulted in many publications and 15 PhD theses being funded from CfG. Also, besides my own position as Professor in gender cardiology, Ivanka Savic-Berglund, one of the researchers from CfG was appointed as Professor in gender neurology at KI. Four other researchers, Juan-Jesus Carrero, Nina Johnston, Ann Fridner, and Karolina Kublickiene, were appointed as associate professors.

Education

In the years since 2008 we changed policy and aimed at creating research positions instead of research funding. Instead, we began more educational efforts.

Together with Dr. Nina Johnston, at that time Educational Leader at the CfG, I made a lecture tour in Ontario, Canada, in 2007. We visited four universities with special activities in gender-specific medicine. At Mac Master University in Hamilton we learnt about the web-based educational tool called "The Gender Lens" and brought it home for introduction at KI. The Stockholm County Council gave us the mandate to develop education in gender medicine and since then we had a yearly funding from the County. In 2010 we succeeded in introducing mandatory inclusion of gender medicine in the undergraduate curriculum for medical, nursing, and physiotherapy students at KI. They have half a day during the first year, always with a mandatory examination. We also have lectures in gender medicine at the end of the medical student's education, term 11. However, we have also lectures for the teachers in most of the other educational areas. In 2010 we created a PhD postgraduate course during 1 week every year named Sex and Gender in Biomedical Research. In the same year we introduced the first European web-based postgraduate course in gender medicine called Sex and Gender in Health and Disease. It runs during one term and ends with an examination and is very popular [26].

During 2010 and 2011 we were one of the six participating European universities in the EU-funded "Master Course in Gender Medicine, EU-GIM" with the Institute for Gender in Medicine at the Charité University in Berlin as the organizing center. Several lecturers from CfG were involved in weekend courses as well as two summer courses in Germany and in Italy. Also, course exchange has taken part between CfG and Prof. Alexandra Kautsky-Willer at Gender Medicine Department, University of Vienna and Allgemeiner Kranken Haus (AKH) in Vienna. Educational exchange has also taken place in 2014, 2015, and 2016 between CfG and the National Defense Medical Center at TRI STAR Hospital, Taipee and Taichung Armed Forces General Hospital, Taichung, both in Taiwan where I and Dr. Mia von Euler have been lecturers. In 2017 we began a PhD course named Sex and Gender in Cardiovascular Research with a focus on cardiology, stroke, and vascular surgery at KI. Also since 2007 the center participates in NIH's project in creating an educational platform of Gender Medicine.

Over the years, CfG has published several textbooks. A Swedish textbook about CVD in women was published in 2003 and its third edition was published in 2017 with myself as the editor: KSG [27]. An international textbook about gender differences in neurology

Figure 12.1 Private Photo.

(Editor: Prof. Ivanka Savic-Berglund) was published in 2011 [28] and an international textbook named "Handbook of Clinical Gender Medicine" (Editor-in-Chief: KSG) (Fig. 12.1) was published in 2012 [29]. Also several books for the lay public were written and within the area of gender medicine, several book chapters and review articles as well as research articles have been published [30–31].

NATIONAL AND INTERNATIONAL ACTIVITIES

In May 2002 we had our first international symposium with, among others, invited guests Prof. Marianne Legato of Columbia University, New York and Prof. Judith La Rosa, the first leader of the Office for Women's Health at the National Institute of Health, Washington, DC. Every year thereafter international symposia have taken place at CfG. In May 2005 a Wallenberg symposium was arranged on brain differences in men and women with speakers from the United Kingdom and the United States. In 2008 CfG was the organizer of the third International Society of Gender Medicine (IGM) Congress in Stockholm and about 400 participants from the whole world were gathered, including individuals from Japan and Australia. I was one of the co-founders of the International Society of Gender-Medicine (IGM) At the first symposium of gender medicine in Berlin in 2006, the foundation of an International Society of Gender Medicine was proposed. The organization was then established in 2007 in Berlin with Prof. Vera Regitz-Zagrosek and Prof. Karin Schenck-Gustafsson as signers of the official legal documents. These two cardiology professors were also the first presidents of IGM. Since 2009 Prof. Marek Glezerman from Israel has served as president. The following cities have hosted our international congresses: The first congress 2006 was followed by Vienna 2007, Stockholm 2008, Berlin 2009, Tel Aviv 2010, joint meeting with OSSD (Organization of the Study of Sex Differences) Baltimore 2012, Berlin 2015, and Sendai 2017.

A Memorandum of Understanding (MOU) between CfG and the Center for Gender Specific Medicine at Columbia University, New York, NY, was signed as early as 2004. It resulted in two visiting researchers: one from Sweden went to the Department of Cardiology, Medical Center, Columbia University and the other from New York spent a summer at the Karolinska University Hospital in Stockholm. In 2015 another MOU was signed with Food and Drug Administration (FDA), Woman's Office, Washington DC, establishing the relevance of sex and gender aspects to the use of medicines, and beginning the formation of a database that we are building up in this area. In 2012 CfG joined the EU project EU-GenMed, Gender Medicine in Europe, a collaboration between EU institutions both in clinics, preclinics, and education under the lead of Prof. Vera Regitz-Zagrozek. It has resulted in joint publications [32] and finished 2014. It has been followed by the EU project GenCAD with focus on CVD.

RESEARCH

The main focus has been on cardiovascular disease, endocrinology including ob/gyn, neuropsychiatry including stress research, inflammation, and autoimmune diseases with renal diseases.

Among other things we have researched MI without coronary artery obstruction, undertreatment of women with CVD, diagnostics in women, and the role of sex horrmones and CVD. In neurology we have done research about pet-scan of the brains in homo- and heterosexual men and women when exposed to different odors have been investigated. In social science we have explored the working conditions for female and male doctors and found that women doctors more often are exposed to harassment and have more depression and sick leave time [33,34]. One ongoing project is the introduction of a subscription support ("Gender Button") for doctors when prescribing medications to patients. We are building a unique data bank of sex and gender evidence based drug utilization together with the Swedish Medical Product Agency and the American FDA (janusmed.sex and gender). The aim is to give the right dose and the right drug at the right time for both men and women [35–38].

REFERENCES

[1] Al-Khalili F, Svane B, Di Mario C, Ryden L, Schenck-Gustafsson K. Intracoronary ultrasound measurements in women with myocardial infarction without significant coronary lesions. Coron Artery Dis 2000;11:579–84.

[2] Horsten M, Mittleman MA, Wamala SP, Schenck-Gustafsson K, Orth-Gomér K. Social relations and the metabolic syndrome in middle-aged Swedish women. J Cardiovasc Risk 1999;6:391–7.

[3] Horsten M, Ericson M, Perski A, Wamala SP, Schenck-Gustafsson K, Orth-Gomér K. Psychosocial factors and heart rate variability in healthy women. Psychosom Med 1999;61:49–57.

[4] Eriksson M, Egberg N, Wamala S, Mittleman MA, Schenck-Gustafsson K. Relationship between plasma fibrinogen and coronary heart disease in women. Arterioscler Thromb Vasc Biol 1999;19:67–72.

[5] Orth-Gomer K, Horsten M, Wamala SP, Mittleman MA, Kirkeeide R, Svane B, et al. Social relations and extent and severity of coronary artery disease. The Stockholm Female Coronary Risk Study. Eur Heart J 1998;19:1648–56.

[6] Wamala SP, Mittleman MA, Horsten M, Schenck-Gustafsson K, Orth-Gomér K. Job stress and the occupational gradient in coronary heart disease risk in women. The Stockholm Female Coronary Risk Study. Soc Sci Med 2000;51:481–9.

[7] Orth-Gomer K, Wamala SP, Horsten M, Mittleman MA, Kirkeeide R, Svane B, et al. Marital stress worsens prognosis in women with coronary heart disease: The Stockholm Female Coronary Risk Study. JAMA 2000;284:3008–14.

[8] Horsten M, Mittleman MA, Wamala SP, Schenck-Gustafsson K, Orth-Gomér K. Depressive symptoms and lack of social integration in relation to prognosis of CHD in middle-aged women. The Stockholm Female Coronary Risk Study. Eur Heart J 2000;21:1072–80.

[9] Al-Khalili F, Wamala SP, Orth-Gomer K, Schenck-Gustafsson K. Prognostic value of exercise testing in women after acute coronary syndromes (The Stockholm Female Coronary Risk Study). Am J Cardiol 2000;86:211–13.

[10] Al-Khalili F, Svane B, Wamala SP, Orth-Gomér K, Rydén L, Schenck-Gustafsson K. Clinical importance of risk factors and exercise testing for prediction of significant coronary artery stenosis in women recovering from unstable coronary artery disease: the Stockholm Female Coronary Risk Study. Am Heart J 2000;139:971–8.

[11] Wamala SP, Mittleman MA, Horsten M, Eriksson M, Schenck-Gustafsson K, Hamsten A, et al. Socioeconomic status and determinants of hemostatic function in healthy women. Arterioscler Thromb Vasc Biol 1999;19:485–92.

[12] Wamala SP, Mittleman MA, Schenck-Gustafsson K, Orth-Gomér K. Potential explanations for the educational gradient in coronary heart disease: a population-based case-control study of Swedish women. Am J Public Health 1999;89:315–21.

[13] Wamala SP, Mittleman MA, Horsten M, Schenck-Gustafsson K, Orth-Gomér K. Short stature and prognosis of coronary heart disease in women. J Intern Med 1999;245:557–63.

[14] Wamala SP, Lynch J, Horsten M, Mittleman MA, Schenck-Gustafsson K, Orth-Gomér K. Education and the metabolic syndrome in women. Diabetes Care 1999;22:1999–2003.

[15] Eriksson-Berg M, Silveira A, Orth-Gomer K, Hamsten A, Schenck-Gustafsson K. Coagulation factor VII in middle-aged women with and without coronary heart disease. Thromb Haemost 2001;85:787–92.

[16] Koertge J, Al-Khalili F, Ahnve S, Janszky I, Svane B, Schenck-Gustafsson K. Cortisol and vital exhaustion in relation to significant coronary artery stenosis in middle-aged women with acute coronary syndrome. Psychoneuroendocrinology 2002;27:893–906.

[17] Eriksson-Berg M, Egberg N, Eksborg S, Schenck-Gustafsson K. Retained fibrinolytic response and no coagulation activation after acute physical exercise in middle-aged women with previous myocardial infarction. Thromb Res 2002;105:481–6.

[18] Al-Khalili F, Svane B, Janszky I, Rydén L, Orth-Gomér K, Schenck-Gustafsson K. Significant predictors of poor prognosis in women aged ≤65 years hospitalized for an acute coronary event. J Intern Med 2002;252:561–9.

[19] Weidner G, Kohlmann CW, Horsten M, Wamala SP, Schenck-Gustafsson K, Högbom M, et al. Cardiovascular reactivity to mental stress in the Stockholm Female Coronary Risk Study. Psychosom Med 2001;63:917–24.

[20] Koertge JC, Ahnve S, Schenck-Gustafsson K, Orth-Gomer K, Wamala SP. Vital exhaustion in relation to lifestyle and lipid profile in healthy women. Int J Behav Med 2003;10:44–55.

[21] Janszky I, Mukamal KJ, Orth-Gomer K, Romelsjö A, Schenck-Gustafsson K, Svane B, et al. Alcohol consumption and coronary atherosclerosis progression—the Stockholm Female Coronary Risk Angiographic Study. Atherosclerosis 2004;176:311–19.

[22] Janszky I, Ericson M, Mittleman MA, Wamala S, Al-Khalili F, Schenck-Gustafsson K, et al. Heart rate variability in long-term risk assessment in middle-aged women with coronary heart disease: The Stockholm Female Coronary Risk Study. J Intern Med 2004;255:13–21.

[23] Al-Khalili F, Janszky I, Andersson A, Svane B, Schenck-Gustafsson K. Physical activity and exercise performance predict long-term prognosis in middle-aged women surviving acute coronary syndrome. J Intern Med 2007;261:178–87.

[24] Wang HX, Leineweber C, Kirkeeide R, Svane B, Schenck-Gustafsson K, Theorell T, et al. Psychosocial stress and atherosclerosis: family and work stress accelerate progression of coronary disease in women. The Stockholm Female Coronary Angiography Study. J Intern Med 2007;261:245–54.

[25] Nagy E, Janszky I, Eriksson-Berg M, Al-Khalili F, Schenck-Gustafsson K. The effects of exercise capacity and sedentary lifestyle on haemostasis among middle-aged women with coronary heart disease. Thromb Haemost 2008;100:899–904.

[26] Miller VM, Kararigas G, Seeland U, Regitz-Zagrosek V, Kublickiene K, Einstein G, et al. Integrating topics of sex and gender into medical curricula-lessons from the international community. Biol Sex Differ 14 October 2016;7(Suppl 1):44. e Collection 2016.

[27] Schenck-Gustafsson K, Johnston N. Kvinnohjärtan, Hjärt-kärlsjukdomar hos kvinnor (Women's heart: CVD in women), vol. 3. Lund: Studentlitteratur; 2017. ISBN X.

[28] Savic-Berglund I, volume editor. Differences in the human brain, their underpinnings and implications—progress in brain research, vol. 186. New York: Elsevier Science and Technology. ISBN 978-0-4453-6303.

[29] Schenck-Gustafsson K, DeCola PR, Pfaff DW, Pisetsky DS. Handbook of clinical gender medicine [Editor in chief: Schenck-Gustafsson Karin]. Basel: Karger; 2012. ISBN 978-3-8055-9929-0.

[30] Humphries KH, Izadnegahdar M, Sedlak T, Saw J, Johnston N, Schenck-Gustafsson K, et al. Sex differences in cardiovascular disease—impact on care and outcomes. Front Neuroendocrinol 18 April 2017; pii: S0091-3022(17)30020-1. Available from: http://dx.doi.org/10.1016/j.yfrne.2017.04.001. [Epub ahead of print] Review.

[31] Berglund A, Schenck-Gustafsson K, von Euler M. Sex differences in the presentation of stroke. Maturitas May 2017;99:47–50. Available from: http://dx.doi.org/10.1016/j.maturitas.2017.02.007. Epub 16 February 2017. Review.

[32] The EUGenMed, Cardiovascular Clinical Study Group, Regitz-Zagrosek V, Oertelt-Prigione S, Bossano-Prescott E, Franconi F, et al. Gender in cardio-vascular disease: impact on clinical manifestations, management and outcomes. Eur Heart J January 2016 1;37(1):24–34. Available from: http://dx.doi.org/10.1093/eurheartj/ehv598.

[33] Gustafsson Sendén M, Schenck-Gustafsson K, Fridner A. Gender differences in Reasons for Sickness Presenteeism—a study among GPs in a Swedish health care organization. Ann Occup Environ Med 20 September 2016;28:50. Available from: http://dx.doi.org/10.1186/s40557-016-0136.

[34] Gustafsson Senden M, Lovseth LT, Schenck-Gustafsson K, Fridner A. What makes physicians go to work while sick: a comparative study of sickness presenteeism in four European countries (HOUPE). Swiss Med Wkly 2013;143:w13840.

[35] Karlsson Lind L, von Euler M. Korkmaz S, Schenck Gustafsson K, Sex and gender differences in medicines—a web-based open access knowledge database (in review).

[36] Loikas D, Wettermark B, von Euler M, Bergman U, Schenck-Gustafsson K. Differences in drug utilisation between men and women: a cross-sectional analysis of all dispensed drugs in Sweden. BMJ Open 2013;3:e002378. Available from: http://dx.doi.org/10.1136/.

[37] Loikas D, Karlsson L, von Euler M, Hallgren K, Schenck-Gustafsson K, Bastholm Rahmner P. Does patient's sex influence treatment in primary care? Experiences and expressed knowledge among physicians—a qualitative study. BMC Fam Pract 13 October 2015;16(1):137. Available from: http://dx.doi.org/10.1186/s12875-015-0351-5.

[38] Maas AH, Euler MV, Bongers MY, Rolden HJ, Grutters JP, Ulrich L, et al. Abnormal uterine bleeding in premenopausal women taking oral anticoagulant or antiplatelet therapy. Maturitas 1 September 2015; pii: S0378-5122(15)30049-9. Available from: http://dx.doi.org/10.1016/j.maturitas.2015.08.014.

CHAPTER 13

From Appalachia to West Texas: A Journey to Move Beyond One-Sex Medicine

Marjorie R. Jenkins
Texas Tech University Health Sciences Center, Lubbock, TX, United States

Several life events led to my dedication to focus a career on sex and gender differences in medicine. The first was an exposure to gender as a health-care disparity and occurred when I was a young child. I grew up in Belfry, Kentucky, a region deep in the heart of Appalachian coal country. My mother was fifteen years old and pregnant when she quit school to marry a Navy seaman twelve years her senior. I am the youngest of eight children born to them before she became a widow at age thirty-five.

After my father's death, my mother worked two to three jobs while pursuing her general educational development (GED), and an Associate's Degree in Nursing. Despite her hard work, we were poor, lacking health insurance and access to health care. In retrospect, this experience instilled in me at a young age an intuitive knowledge of health disparities. I recall being ill with high fevers and having the common accidents that befall active children. Mom would take me to see my maternal grandfather, whom I believed to be a doctor because he worked at the hospital. Later, I learned he was forced to leave his job in the coal mines due to pneumoconiosis (black lung disease). Only then he became the hospital janitor.

Learning my grandfather was a janitor did not make me trust him any less, it made me want to become a doctor even more! Imagine my disappointment when a guidance counselor informed me I was too poor to go to medical school. She added insult to injury when she explained that there was no influential person in my life to help me. In other words, I was told to choose a career attainable by someone of my socioeconomic status. Being 16 and worried about future employment, I selected a major in Chemical Engineering rather than Biology

The International Society for Gender Medicine. DOI: http://dx.doi.org/10.1016/B978-0-12-811850-4.00013-2

and premedicine. In retrospect, I realize my childhood development and educational experiences were common stumbling blocks for the socially disadvantaged. However, they would serve me well as the seed which would grow into my passion for societal and gender-health disparities. That passion would become the driving force behind my choice to pursue a career in sex- and gender-based medicine (SGBM).

Another influence on my future in SGBM was my work as a chemical engineer control systems designer for the Eastman Kodak Company. Engineers are notoriously logical, evidence-based, and precise. There is no wiggle room when you are controlling for a 100-ton benzene process that is highly combustible. In that environment, precision was a lifesaving quality. That quality helped me to become a tireless advocate for the development of lifesaving precision medicine through the incorporation of sex and gender differences. But, first, I had to become a doctor! My dream born so long ago refused to die.

After some research, I learned of a relatively new medical school, Quillen College of Medicine at East Tennessee State University located a few miles away from Kodak Eastman. Realizing that most medical school graduates were in debt, and admission is not based on who you knew but what you knew, I applied, and graduated in 1995. I am certain it was my engineering foundation which helped spark an interest in the burgeoning evidence of how sex and gender influence health and disease.

Another major influence was the choice to train in internal medicine instead of obstetrics and gynecology, a choice that would lead me closer to a career centered on SGBM. During my early years of internal medicine residency training, I was exposed to the work of pioneers in sex differences in cardiovascular disease, pioneers such as Dr. Laura Wexler at the University of Cincinnati. I underwent training in an environment which required knowledge of the whole person and the myriad of influences between the various organ systems. For example, the emerging research on the impact of reproductive hormones on the cardiovascular system was astounding.

After completing residency training at the University of Cincinnati, I entered private practice where as an internist I was able to see firsthand, albeit anecdotally, the differences in communication, compliance, and treatment response between men and women. I became determined to learn more about the scientific foundation of these differences, to delve deeper than the simplistic approach of "men are from Mars, women are from Venus." Believing the scientific evidence would provide practitioners with a far better capability to care for both men and women, I began a self-study of the sex differences literature that was beginning to emerge during the late 1990s and early 2000s.

In 2001, the Chair of Internal Medicine at the Texas Tech University Health Sciences Center (TTUHSC), Dr. Donald Wesson, contacted me at the behest of a colleague. An invitation was extended to interview for an academic position that would provide an opportunity to work in both the Departments of Internal Medicine and Obstetrics and Gynecology. Designated as a women's health internist, I would be trained in office gynecology and menopause management, as well as expanding women's health training for Internal Medicine and OB-GYN residents. I recall thinking, "Where is Lubbock, Texas?"

At the urging of my husband who knew of my passion for education, I made a visit to TTUHSC's main campus in Lubbock in late 2001. During this time I reconnected with Dr. Steven Berk, my ETSU medical school advisor and the then Chair of Internal Medicine at ETSU Quillen. His guidance had been a significant influence on my decision to become an internist. At that time, Dr. Berk was TTUHSC regional dean in Amarillo, Texas, located two hours north of Lubbock. On a visit to the Amarillo campus of Texas Tech, I found out that Dr. Berk and other faculty members had founded a Women's Health Research Institute. He was determined to make women's health a signature program for the Amarillo campus and recruited me to Lubbock to join in his work. My journey into sex and gender medicine had begun in earnest.

Upon arrival in Amarillo in June 2001, the community immediately embraced me in the friendliness and hospitality that is the norm in West Texas. During the rounds of community outings such as the symphony, theater, and community teas, I was often asked what type of medicine I practiced. A growing desire led me to reply, "I am focusing my practice on women's health." The most common responses to this statement were, *Can you deliver my baby?* or *Can you perform my hysterectomy*? Henceforth, I would jokingly share that women either wanted to use it or lose it and wanted me to help. However, I did not deliver babies or perform hysterectomies. What I did was practice under a delivery of care model that encompassed addressing seven domains of health: physical, mental, emotional, social, sexual, spiritual, and financial aspects of health, always through a sex and gender lens. From that point forward, when asked what type of doctor I was, I would smile and reply, "I am a womanologist."

In 2002 after one year at TTUHSC, something happened that would alter the course of my medical career from women's health to sex- and gender-specific medicine. Dr. Marianne Legato published *The Principles of Gender-Specific Medicine*, an 800+ page tome of scientific evidence. Being a data and science geek, I read it from cover to cover, walked into Dean Berk's office and told him I was not going to focus on women's health any longer. To this he replied, "But, that is why I hired you!" I set the two volume text on his desk and exclaimed, "This is the wave of the future. It is the best-kept secret in medicine. We are practicing clinical care on a male model and women's health has been relegated to bikini medicine!" Over the next months and years, the conversation was expanded to include men and women, and work toward better health care for all. Although not immediately convinced I was on the right track, Dean Berk did allow me to begin branding our educational programs and clinical practice under sex- and gender-specific women's health. With leadership support, we obtained space and funding for a multidisciplinary Gender-Specific Women's Health Clinic, which opened in 2008.

At the time, many other activities were occurring nationally to progress sex and gender medicine. In 2001 the Institute of Medicine defined sex and gender as unique terms, so from 2002 onward, my practice of medicine through a sex and gender lens became more sophisticated because it was guided by the evolving science. I developed a fellowship program and continued to educate students, residents, and colleagues about the emerging evidence that proved without a doubt that men and women are different even at the cellular level.

The phrase "every cell has a sex" had not caught on by this time, but it would not be long before it became the go-to phrase for sex and gender medicine advocates across the globe. The year 2005 presented the next leg of my professional journey, the opportunity to become the Director of the Amarillo Women's Health Research Institute. As mentioned earlier, the Women's Health Research Institute was founded just before my arrival in Amarillo, but it would play an integral part in TTUHSC's trajectory toward becoming a global force to progress sex and gender within health education, biomedical research, and clinical care.

Having only four years of academic experience under my belt made me hesitant about assuming leadership of the Institute. I had not spent much time conducting research in the field. However, I agreed to codirect the institute with a colleague from our School of Pharmacy. Dr. Margaret Weis and I were an instant dynamic duo. Funded by NIH, she was already a passionate women's health and sex differences researcher. Her passion and my interest in ensuring that the scientific evidence made it to the patient and health care consumer combined for a perfect fit. In 2007 our Midland Texas regional dean, Dr. John Jennings, contacted me. He was based in Odessa, Texas, the hometown of the First Lady Laura W. Bush. He had watched an interview with Mrs. Bush in which she discussed the differences between men and women related to heart disease. Many did not know that Mrs. Bush was the inaugural and current lifelong ambassador for the National Heart Lung and Blood Institute's Heart Truth Campaign. Knowing this, Dr. Jennings wanted to launch a women's health institute under her name, but at the time the Amarillo Women's Health Institute was expanding across the TTUHSC campuses. To propose a Laura W. Bush Institute for Women's Health, we would need the support of Texas Tech Chancellor, Mr. Kent Hance.

Chancellor Hance had a long history with the Bush family. In 1978 he won the seat for the 19th Congressional District, becoming the first and only person ever to defeat George W. Bush in an election. Chancellor Hance often joked if he had lost that election he would have been president of the United States and George W. Bush would be Chancellor of TTU. Nevertheless, he had a close and abiding friendship with George W. Bush and Laura W. Bush that continues today. With this history in mind, Dr. Jennings and I, along with other leaders from TTUHSC, met with the Chancellor. At that meeting, we proposed the idea of requesting First Lady Laura W. Bush to allow us to name our women's health institute as The Laura W. Bush Institute for Women's Health.

Expressing his support, Chancellor Hance charged me with writing the formal proposal. With input from our development and communications staff, I developed the proposal and sent word that it was complete. I received a response that TTU had decided not to proceed with the proposal. I was deeply disappointed given our institution's potential to connect women's health research, education, and outreach across our multiple campuses and 130,000 geographical square miles. I believed that we had an unprecedented opportunity to progress sex- and gender-specific health under the name "Laura W. Bush." Based on this belief, I wrote a note and sent copies of the proposal to Chancellor Hance, TTUHSC Interim President Bernie Mittermeier, and School of Medicine Dean Steve Berk. I was delighted to receive word that the proposal would be delivered to Mrs. Bush via her former Chief of Staff, Ms. Andi Ball of Austin.

In July 2008, we were invited to pitch our proposal to the First Lady of the United States. Soon after, I found myself standing at the security gate of the White House awaiting my escort to the East Wing. It was a bit surreal. As the meeting progressed, I was given the floor and the opportunity to express how clinical care for women was missing the essential perspective of sex and gender. With the exceptions of pregnancy, lactation, and menopause, we were approaching and treating women's health as if there was no difference between the health needs of women and men. This was a myth, as scientific research was revealing crucial differences between the sexes in major organ systems such as the brain and heart at a rapid pace. I spoke about the enormous impact of having the Laura W. Bush name attached to an Institute for Women's Health. The potential positive effect on the national and global movement to bring sex and gender differences science to the forefront of research, education, and clinical care was immeasurable. Interestingly, Chancellor Hance's Chief of Staff, Mr. Jodey Arrington, was with our team at the White House that day. Before joining the Chancellor's team, he had been a member of George W. Bush's staff for over a decade, from the Texas governorship to the White House. Today, as if coming full circle, Congressman Arrington is the freshman congressman for the 19th Congressional District, the same seat won by Congressman Kent Hance in 1978 and remains a staunch supporter of advancing the health of women and families.

Mrs. Bush did gift her name to this multicampus multidisciplinary institute, and the Laura W. Bush Institute for Women's Health, www.laurabushinstitute.org, became a reality. I was appointed to the position of Executive Director and had the privilege of helping shape the institute's mission. I did so around sex and gender differences. It took several years for the Institute to obtain traction and mature into its current laser-focused mission of progressing women's health through research, education, and clinical care. Through the hard work and dedication of many, the support of the generous Amarillo community where it all began, and the support of other amazing Texas communities that followed (Midland, Lubbock, San Angelo, Abilene, and Dallas), the Laura W. Bush Institute became a respected and expected voice in the conversation about sex and gender differences in policy, health professional's education, research, and consumer outreach. I would love to write that this achievement was without naysayers; however, it was not. But, the leadership of TTU and TTUHSC stood firm on the evolution of sex and gender medical science, maintaining the belief that we could contribute to progress sex- and gender-specific health care in West Texas communities and beyond.

In 2009 my career progressed along its gender-health trajectory when I was invited by our Dean to create a sex- and gender-specific undergraduate medicine curriculum. Under the mantel of the Laura W. Bush Institute, and with the support of TTUHSC President Tedd Mitchell, I began my work in this area. After researching the issue nationally and internationally, I realized that stand-alone curricula are costly and often became outdated and forgotten within health professionals' institutions. After several discussions with leadership, faculty, and students, we opted to create adaptable interprofessional sex and gender educational materials that could be utilized in a multitude of ways within schools of pharmacy, medicine, nursing, and allied health. These peer-reviewed educational products can be found at www.sexandgenderhealth.org and are available via open access due to the generosity of our Laura W. Bush Institute National Advisory Board. The products include interactive modules, slides sets, case simulations, and podcasts. An unexpected but valuable outcome of this project was the realization that there was a major barrier to the achievement of curricular integration across US medical schools.

In 2013 while mulling over the idea of hosting a national educational summit, Dr. Eliza Chin, Executive Director of the American Medical Women's Association, contacted me to discuss partnering to host just such a summit. In 2015 the Mayo Clinic, the Laura W. Bush Institute for Women's Health, the American Medical Women's Health Association, and the Society for Women's Health Research were the premier sponsors of the 2015 US Sex and Gender in Medical Education Summit. One hundred and thirteen US, Canadian, and European Medical Schools were represented at the summit. Information about the 2015 Sex and Gender Medical Education

Summit and its successor, the 2018 Sex and Gender in Health Professions Education Summit, can be found at sghesummit2018.com. The 2018 Summit will welcome over 300 health professionals from medicine, nursing, pharmacy, and dental and allied health to the University of Utah School of Medicine.

In 2015 after fourteen years of experience in sex and gender medicine within academia, I was invited to join the US Food and Drug Administration's Office of Women's Health as the Director of Medical Initiatives and Scientific Engagement. My time with the Office of Women's Health has allowed me an opportunity to learn from within the complexities of subpopulation analysis and development of national health policy. Also, it has afforded the opportunity to develop sex and gender professional education programs internal and external to FDA. Currently, I am working with the FDA's Office of Women's Health, and continuing health professional education work as the Chief Science Officer for the Laura W. Bush Institute for Women's Health.

My professional journey has been exciting and challenging, and each stage has brought me closer to achieving my goal of shining a bright light on the differences which exist in sex and gender health. I am committed to ensuring that women and men, boys and girls, and those across the gender spectrum receive evidence-based precision health care. As you have read within other chapters of the book, we have made outstanding strides in advancing sex as a biologic variable in biomedical research, integrating scientific discoveries into health professional's education, and educating consumers to be their own health advocates by seeking those providers who "know the difference."

In conclusion, I offer two questions, which continue to spur me along when I hit barriers that seem insurmountable. First, how can we achieve personalized medicine if we ignore sex and gender, two health variables that every person has without fail? Second, how can a scientific discovery save a life if it is not shared, through education, with those who can apply it to patient care? My answer to both is that neither can be achieved without the continued progress of sex and gender health and this is why the journey continues. We are farther along than we were but not there quite yet.

CHAPTER 14

Sex and Gender in Emergency Medicine: Advancing Care Through Person-Specific Research, Education, and Advocacy

Alyson J. McGregor
Brown University, Providence, RI, United States

SEX AND GENDER IN EMERGENCY MEDICINE

I first began to lecture about sex and gender in Emergency Medicine (EM) in 2005; I was a new attending physician and many of my professional associates expressed their reluctance to embrace a new viewpoint of medicine. It turns out that I was ahead of the curve, as many people liked to point out. I was on the right track, at the very least. The ideas that I expressed in that first lecture were big, and important enough for fellow physicians to take notice. The title of the talk was simply "Gender." I was speaking to my EM Department about the historical lack of research focused on women and the consequences of that deficiency. I reviewed the ways in which this lack contributed to our understanding and misunderstanding of the best ways to care for acutely ill patients in the emergency department (ED). In retrospect I regard this lecture as my "Coming Out Party." I was beginning my academic career by declaring to the world that sex and gender would be my area of study and expertise and defining those important differences between women and men in health and disease might make a huge difference in treatment and treatment outcomes.

Throughout my residency training, I would mention to colleagues that I was interested in "women's health," immediately provoking my associates to assume that I intended to focus on obstetrical and gynecological emergencies and their treatment. My fellow residents would often seek me out as the resident expert, claiming that I should perform any needed pelvic exams. Clearly I was viewed as somehow more qualified to perform this routine exam element because I cared about women. Well, they were partially correct. I did and do care about the

The International Society for Gender Medicine. DOI: http://dx.doi.org/10.1016/B978-0-12-811850-4.00014-4

health of women...the entire bodies and minds of women and how their unique anatomy and physiology controls their health from every perspective, affecting every part of their bodies, not just the "private" areas. I began to address the medical, political, social, and psychological conundrum: women's health was inescapably linked in the minds of health care providers with reproduction and only reproduction. All other medical care of women was considered to be beside the point, or irrelevant, that is, what was good for the goose was good for the gander irrespective of gender (or sex).

I began to read widely in this field seeking to expand my fund of knowledge so that I could educate. I began with the seminal, and, in fact, only definitive textbook, one that would inspire and propel me forward providing education, ammunition, direction, and the passion and perseverance I would need to advance within this field. The book, *Principles of Gender-Specific Medicine* by Marianne Legato, provided the evidence, the proof that I had been seeking that men and women were different and providing the same diagnostic testing, therapeutic interventions, pharmaceuticals, and procedures to both groups in equal measure invoked within me a resounding cognitive dissonance and a resolve to enlighten providers within my specialty [1].

The Society for Academic Emergency Medicine (SAEM) holds its annual scientific conference in the spring. Shortly after my "Gender" lecture in 2005, I prepared a didactic session for SAEM entitled "Women's Health and Gender-Specific Research in Emergency Medicine: Yesterday's Neglect, Tomorrow's Opportunities," which included a panel of experts and focused medical content. I was certain that this session would be the turning point for EM practitioners. My colleagues would finally understand and embrace the idea that had become self-evident to me: we had to treat the men and women in our practice differently, taking into account the fact that they were not the same. My team assembled in the conference room, uploading our slides, and gearing up. But, instead of the pivotal moment I envisioned...we looked out to an empty room. There was no audience. Not one person in the scientific community of EM believed that Sex and Gender Medicine had any relevance for their research or practice. I was in despair; how could I spread the good news when no one was willing to listen to the message?

In an effort to seek out like-minded individuals, the following year I headed to the American Medical Women's Association annual conference. I was drawn to an impromptu gathering being advertised with hand-made signs. "Women's Health Initiative Committee" was meeting in Dr. Jan Werbinski's hotel room. There were 10 of us in the room that day, sharing our vision of where we thought women's health needed to be. A number of the women focused on what was required to improve reproductive care, but I initiated a discussion of how highlighting and exploring sex and gender differences was a path to truly improve women's health. Dr. Werbinski and I felt a similar passion; she introduced me to Dr. Marjorie Jenkins who was the Director and Chief Scientific Officer for the Laura W. Bush Institute of Health at the time. Dr. Jenkins connected me to her growing national network. This group collectively developed an organization designed to increase the awareness of sex and gender in medical education called the Sex and Gender Women's Health Collaborative (SGWHC).

Over the past 10 years, this collective group of interdisciplinary sex- and gender-based health champions has shared an understanding of the real challenges facing the acceptance and integration of sex and gender into medical research, education, and practice. This network, constantly growing, reaching out, and connecting with new members, has created momentum, supporting research and advancing our understanding of the impact of sex and gender on health while continuing to lobby for the inclusion of these novel ideas into medical education.

My growing relationship with this national network of leaders in the field of sex and gender medicine provided the support and encouragement that I needed to address, again, colleagues in my own specialty of EM. EM practitioners treat more than 100 million Americans in ED's every year. My goal was to raise the awareness of every single EM health care practitioner who treated each one of these patients. Despite my initial feelings of frustration because of poor attendance during our first attempt at a didactic session on sex and gender, I realized that the opportunity and potential to create outreach. I discovered a like-minded partner in Dr. Esther Choo. We cofounded "Women's Health in Emergency Care" (WHEC) in 2010, which is a program supporting research and education in sex and gender medicine for EM physicians. Interest in the field grew quickly, but as "women's health experts" we continued to be seen as specialists in "reproductive medicine." Every conversation with students, faculty, and collaborators locally and nationally began with the necessary explanation that "women's health" meant all the ways in which women were different in body and pathology than men. It also became obvious to us, that unless we changed our name, we would never have enough men seated at the table to create real change.

By 2012 the term "sex" began to be used in the medical and scientific realms to denote the biological construct; this sea change in perspective was local, national, and federal as demonstrated by the establishment of Specialized Centers of Research on Sex Differences Awards by the Office of Women's Health Research. Officially "sex" was different from "gender" and both terms held specific meaning when referenced in research and clinical practice. We then purposefully chose to change the name of our program from WHEC to Sex and Gender in Emergency Medicine (SGEM). This decision was politically charged and required extensive discussions with board members, department leadership, national advisors, and community leaders. Many concerns were voiced: Would people understand and accept the terminology? What would happen to our supporters who resonated with the term "women's health"? Was it taboo to use the term "sex" in an academic title? Ultimately, we felt that the advantages, such as clarity of our mission and positively influencing the field of sex- and gender-based medicine, would outweigh any potential discomfort in transitioning.

As our program grew, we successfully researched sex- and gender-based factors in chest pain, sepsis, substance use, violence, and stroke. We developed educational programs for medical students and residents, and established a novel 2-year fellowship program with options for research or clinical education; we published a curriculum that included core competencies. We applied for and were granted the status of a Division within our Department of EM, with increased access to departmental funding and influence.

In an effort to include the larger voice of our community in the model for our Division, we established a SGEM Community Advisory Board to bolster our understanding of how sex and gender affect the way health care is received by the patient and the community at large. With the input of our Community Board, we created a patient awareness campaign placing posters in care areas of the ED designed to educate patients about their symptoms and treatment from the perspective of sex and gender. The posters all carry the tag line, "Your Emergency

Is As Unique As You Are. We Know The Difference." This campaign was linked with faculty, student, nursing, and staff education with nine videos each 9 minutes in length focusing on the sex and gender issues most importantly associated with commonly treated ED conditions like trauma, sports injuries, cardiovascular disease, neurovascular emergencies, pain, violence, and substance abuse.

As the SGEM Division grew, we created strong relationships with national EM collaborators who were also passionate about including sex and gender in EM. Drs. Basmah Safdar and Marna Greenberg joined Dr. Choo and me to form the Executive Steering Committee for the 2014 SAEM Consensus Conference: "Gender-Specific Research in Emergency Care," a day-long program designed to set the national research agenda on how sex and gender impact EM research, education, and clinical care. For 2 years we worked obtaining funding, identifying interested researchers, creating a successful and meaningful agenda, and publishing the proceedings; the outcome was well worth the time and effort.

On the day of the conference, I could not help but recall the moment several years earlier when I stood before this same national organization to discuss this very issue; at that moment I had addressed an empty room. This time however, the space was filled with more than 100 EM practitioners including leaders in their research fields. The discussions were lively; challenges and barriers were recognized; a definitive research agenda was identified. Clearly, our specialty of EM was ready to join the national movement to advance sex- and gender-based health care education, practice, and research.

One important lesson I have learned on this journey, and in organizing, promoting, and supervising these events is that by identifying key people who successfully research important areas in EM, and convincing them to review their field through the sex and gender lens, is that these same people have now joined our crusade. Many of them have become experts in sex and gender! These faculties who became involved in the consensus conference grew to have a unique and in depth understanding of the impact of sex and gender on their particular area of study...a perspective most had not previously considered. This community within EM now contributes to many of the projects SGEM has taken on, including the publication of a medical textbook *Sex and Gender in Acute Care Medicine*, the first textbook of its kind focused on sex and gender in the evaluation and treatment of patients in the delivery of acute medical care [2]. I was proud to see this book comc to life in spite of the considerable amount of effort and collaborative work that it required, especially among my coeditors Drs. Esther Choo and Bruce Becker.

The specialty of EM has a unique standing within medicine, in that it is multidisciplinary and has access to an ever-increasing patient population that presents with acute illnesses and injuries that are associated with high morbidity and mortality; in the ED, practicing state of the art sex and gender medicine can have life-saving impact. We still have many goals to achieve working to translate sex- and gender-based medicine into the specialty of EM. We would like to see other EM departments across the country establish similar Divisions of SGEM utilizing the structure and plan of our work promoting sex- and gender-based research, education, and advocacy and making it even better. Recently we established a national SGEM Interest Group through SAEM to allow continued collaboration and growth. Through this effort, new educational tools are being developed to share with EM residency departments, which include the very necessary component of faculty development.

Resistance to change can feel like an overwhelming obstacle. I am more optimistic now. My experience speaking with colleagues in Departments of EM across the country has broadened my understanding of the need for raising EM physicians' consciousness about this issue. Most researchers, educators, and practitioners want to stay current, educated, and effective. Once I have created the "ah-ha" moment on the importance of considering patient sex and gender, they always ask, "What should I do?" Answering this question is often difficult. We in the medical community are participating in a paradigm shift; in a cognitive transformation. We are learning how to establish patients' biological sex and current gender identity, integrating this information into our understanding of the presentation of their disease, risk assessment, proper testing, effective treatment, and communication strategies. We must continue to create momentum by establishing steering committees, interprofessional stakeholders, integrated residency education, updated clinical practice guidelines; we must keep our eye on the future generation by empowering new SGEM leaders through residency training, fellowship programs, faculty development, and local and national research funding initiatives. Change comes in millimeters; evolution, inevitable.

The ultimate goal for EM, and for the practice of medicine in general, however, should be that special divisions of SGEM are no longer necessary. If every EM researcher was able and willing to integrate sex- and gender-based research into their research programs, every EM educator was able to include sex- and gender-based curricular components into their educational training programs, and every EM physician was able to understand sex and gender as human attributes, which must be integrated into every aspect of the clinical care that they provide, the research that they carry out, their very perspective in seeing another human being, whether patient, research participant, or colleague, then, and only then, SGEM programs will become redundant and unnecessary. We welcome that day.

ACKNOWLEDGMENT

I thank Dr. Bruce Becker for editing this chapter.

REFERENCES

[1] Legato M. Principles of Gender-Specific Medicine. 1st ed. San Diego, CA: Academic Press; 2004.

[2] McGregor AJ, Choo E, Becker B. Sex and Gender in Acute Care Medicine. New York, NY: Cambridge University Press; 2016.

CHAPTER 15

A Comparative Physiologist's Journey Into the Field of Sex- and Gender-Specific Medicine

Virginia M. Miller
Mayo Clinic, Rochester, MN, United States

INTRODUCTION

My journey into the field of sex- and gender-specific medicine was anything but deliberate. Having completed a Bachelor's of Science degree in biology education from Slippery Rock State Teacher's College (now Slippery Rock University, Slippery Rock, PA), I thought I was destined to teach high school biology. However, as many other WWII baby-boomers held similar career expectations, teaching jobs were hard to find. At the suggestion of one of my undergraduate professors, Gilbert Dryden, PhD, applying to graduate school seemed like a reasonable option. I was accepted into the Department of Veterinary Medicine graduate school at the University of Missouri. It was the 1970s. While the women's movement was in full swing, science was male dominated. I was the only female graduate student in many of my classes and the environment was not always welcoming or inclusive. Indeed it was not uncommon for *Playboy* pin up pictures to suddenly appear in professional seminars and lectures as "attention-getters." My thesis Professor, Frank E. South, PhD, however, was very supportive. He held a joint position in the Space Science Research Center (now the Dalton Cardiovascular Research Center) where I began studying temperature regulation in hibernating animals. In the early days of space travel, along the lines of *2001: Space Odyssey*, one idea was to isolate a hormone that would lower metabolism in humans, similar to what occurs when animals hibernate, for long space voyages. As only men were astronauts at that time, there was no consideration of women's health or sex differences in response to space travel. The "male norm" characterized medical and physiological

The International Society for Gender Medicine. DOI: http://dx.doi.org/10.1016/B978-0-12-811850-4.00015-6

research. Apart for reproduction, the "70 kg man" was studied as representing all humans. Consideration of the influence of sex on the response, e.g., to zero gravity would not change until 1983 when Sally K. Ride became the first American women to journey into space.

My graduate studies focused on comparing physiological responses among vertebrate animal species and on mammalian adaptations to extreme environments: altitude, deep sea diving, extreme temperatures, and water deprivation. Creative approaches to problem solving were encouraged, such as how to simulate sleep in space on earth. One solution was to construct a sea water tank big enough to house two California sea lions in the heartland of Missouri in order to study their sleep (as related to weightlessness in space). These collective experiences, unbeknownst to me at the time, laid a solid and broad foundation in integrative physiology that has been critical to developing sex- and gender-based research programs. However, at that time the sex of the experimental material was not something we thought about, and as a female graduate student, I deliberately stayed away from anything female related, in order to be taken "seriously" as a future scientist.

Upon completion of graduate school, I followed my husband's career moves, which allowed me to secure postdoctoral positions in neurobiology at the University of Virginia, in cardiovascular physiology at the University of Delaware, and in cardiovascular pharmacology at the Mayo Clinic, Rochester, Minnesota. There I was hired as a research fellow by Paul M. Vanhoutte, MD, PhD, specifically because of my experience in whole animal physiology. It is now the early 1980s, and yes, I was still the only woman research fellow in the laboratory—but not for long! Times were changing. The regulatory properties of the vascular endothelium had just been discovered by Robert F. Furchgott, PhD. Everyone in that laboratory was busy evaluating which chemical and physical stimuli would release endothelium-derived relaxing factor (or EDRF) from these cells. Emerging observational and epidemiological data suggested that noncontraceptive use of estrogen reduced cardiovascular mortality in women [1]. Therefore it was reasonable to hypothesize that estrogen might release EDRF. Thanks to my training in whole animal physiology, another research fellow, Veronique Gisclard, PhD, and I tested this hypothesis. We oophorectomized female rabbits, treated some with intramuscular injections of 17β estradiol and others with placebo. After four days of treatment, we removed the femoral arteries to examine the endothelium-dependent relaxations.

These relaxations were greater in the arteries from the animals treated with estrogen compared to those treated with placebo [2]. This study was the first to identify a mechanism by which a sex-steroid modulated a vascular response. It marked a turning point in my career by confirming the need to include female animals in cardiovascular research. Dr. Furchgott identified EDRF as nitric oxide (NO) in 1986 at the symposium "Mechanisms of Vasodilatation," in Rochester, MN. He received the Nobel Prize in Physiology or Medicine in 1998 for his discovery of NO. I had the pleasure of knowing Dr. Furchgott who wrote a letter of support for my first funded research award, a National Institutes of Health (NIH) R29.

TRAIL BLAZERS

Several events on the national level in the United States stimulated the emerging field of women's health and sex- and gender-based medicine. Florence Hazeltine, MD, PhD encouraged more research into women's diseases at NIH and brought together scientists and other advocates to found the Society of Women's Health Research in 1990. In that same year the Office of Research on Women's Health (Vivian W. Pinn, MD was named as the first director in 1993) was established by the NIH in response to congressional, scientific, and public advocacy concerns about the absence of women in clinical trials. Under the leadership of Bernadine P. Healy, MD[1]. The first woman director of the National Institutes of Health, a large scale, randomized, prospective clinical trial, the Women's Health Initiative (WHI), began in 1991. The WHI investigated whether menopausal hormone therapy reduced cardiovascular disease in women. In addition, she announced a funding opportunity to investigate how estrogen affected arterial function. With colleagues from Mayo (Lorraine A. Fitzpatrick, MD, and Gary C. Sieck, PhD), we successfully secured one of these grants, which allowed us to integrate our research into the national agenda. As a result of this grant, we identified endothelium-derived factors other than NO that were modulated by estrogen and identified sex differences in proliferation of vascular smooth muscle [3–6]. These mechanisms help to explain why there are sex differences in development of cardiovascular disease.

[1]One of the highlights of my career was to receive the Bernadine Healy Award for Visionary Leadership in Women's Health from the Academy of Women's Health in 2014.

Other leaders bringing women's health issues to the national forefront included Marianne J. Legato, MD who published the book, "The Female Heart: The Truth About Women and Coronary Artery Disease" in 1992 and who is the editor of this book (see her own account in Chapter 1), Phyllis Greenberger, MWS, was named the first Executive Director for the Society of Women's Health Research in 1993. Advocacy efforts by these and others brought about the Government Accounting Office report that lead to the passage of legislation, the 1993 NIH Revitalization Act, requiring inclusion of women in clinical trials funded by the NIH.

PROGRESS IN SPITE OF ROCKY ROADS

During the 1990s, the NIH Office of Research on Women's Health also released a call for applications to develop Centers of Excellence in Women's Health. A group of clinicians and scientists at Mayo Clinic began to explore the possibility of applying for this opportunity. Although our proposal to become a designated Center of Excellence in Women's Health was not successful, our efforts provided the framework for establishing a Mayo Clinic Office of Women's Health in 2001. I was named director of the Office, with a primary role to provide education and to expand a research portfolio in topics of women's health. During this time, my laboratory continued work on hormones and sex differences in vascular and platelet functions [7–10]. I also organized a conference sponsored by the American Physiological Society called "Genome and Hormones: An Integrated Approach to Gender Differences in Physiology" (the proceedings of this conference were published in 2004 as *Principles of Sex-Based Differences in Physiology* [11]).

In the same year (2001), The Institute of Medicine published the landmark report "Exploring the Biological Contributions to Human Health: Does Sex Matter?" which concluded that sex matters from womb to tomb in ways that are yet to be considered [12]. This publication confirmed the field of sex- and gender-specific medicine and defined a research agenda for the future.

However, in 2002 the WHI was stopped for an unexpected number of breast cancers, heart attacks, and strokes. The world of hormone research came to a screeching halt as the results of that study seemed to negate previous data on how estrogens benefited vascular health. Although many investigators criticized the design and interpretation of the WHI [13], funding for other hormone-related studies through the NIH became impossible. Around this same time, I learned by attending the meeting of the American Heart Association that a new multicenter clinical trial (the Kronos Early Estrogen Prevention Study or KEEPS) was being designed. This future trial, funded by a private foundation, would examine effects of menopausal hormone treatments on cardiovascular function in menopausal women who were 10 years younger than women included in the WHI. This age group of women was more representative of women included in the observational studies. I thought that having Mayo participate in this trial would be a way for our office to increase our women's health research portfolio while simultaneously giving us a national presence. I contacted the principal investigator, S. Mitchell Harman, MD and secured funding for Mayo as one of the nine study sites. When the study began in 2004 I entered the world of clinical trials.

Although I considered the securing the KEEPS a major success of my tenure as Director of the Mayo Clinic Office of Women's Health, my Department Chairs did not agree. As it was not funded by NIH but by a private foundation, I was told I would lose my job if I did not secure NIH funding! Fortunately we developed several ancillary NIH-funded studies to the main KEEPS protocol at Mayo [14–29]. I kept my job and investigators involved in these ancillary studies formed the backbone of faculty that allowed us to subsequently obtain the NIH training program Building Interdisciplinary Careers in Women's Health and for a Specialized Center of Research on Sex Differences [30].

I became a member of the Board of Directors for the Society of Women's Health Research, and involved in a new organization founded by the Society in 2006, the Organization for the Study of Sex Differences (OSSD) where I served as president from 2010 to 2012. My career in advocacy began with during my tenure as President of OSSD with promotion for inclusion of female animals in basic research studies [31–35]. These efforts supported the NIH requirement for sex to be considered as a biological variable in research applications beginning in 2016. As I write this chapter, it must seem ridiculous to the general reader that it is scientists who have to be told that sex *is* an important biological variable. The notion of the "typical 70 kg man" is hard to erase! It was through the OSSD that I became a member of the board of the International Society of Gender Medicine, an organization that has taken the advocacy of sex and gender differences in health and disease to the international arena. These activities naturally built on each other through engagement and networking. What started out as opportunities for new areas of research through funding opportunities, turned into advocacy and a passion to change the way science is conducted.

NOT THERE YET

These activities are exciting and rewarding. Alternative funding mechanisms to NIH are now considered appropriate career metrics by my Department Chairs. More women are entering scientific careers. However, science and medical leadership remain male dominated and attitudes are hard to change. While *Playboy* pictures are no longer considered appropriate for scientific lectures, biases remain against women scientists and topics related to women's health, which impedes progress. The concept of women's health and sex- and gender-based medicine is politically charged with a focus on reproduction rather than on general health across the life span. The aftermath of the WHI lingers, and physicians and scientists continue to generalize and misinterpret the results of that study; thus, hindering hormone research and confusing women on whether the use of estrogen-related products is safe. Professional societies have increased topics of sex and gender in their annual meetings. However, many scientists are resistant to changing the way experiments are designed and interpreted even when presented with data that the sex of the experimental material will affect their results. Many medical and scientific journals fail to require reporting of data stratified by sex [36,37]. Graduate, medical, and interprofessional health-related educational programs often do not include basic information about sex differences in physiology and pathophysiology [38,39] and thus struggle to integrate these concepts into existing curricula.

FULL CIRCLE?

Fast forward 40 years. Leaving high school biology education behind, my background is being put to good use to educate my colleagues, future researchers, and clinical scientists on topics of integrated sex-based physiology, assessing curricula and programs within my institution, and sharing these ideas with international leaders [40,41]. My diverse experience in comparative and integrated physiology has allowed me to appreciate and seek a board spectrum of topics in which to nurture the next generation of researchers in women's health and sex-based medicine. They are the future of the discipline and there are many topics waiting to be explored. I did not expect to be doing what I am doing. In retrospect, I see that the Lord provided me opportunities and ability to influence science policy. While my circumstances differ from theirs, the words of Queen Esther's uncle Mordechai resonate with me. All of us may be placed and offered opportunities to act "for such a time as this" (Esther 4:14).

ACKNOWLEDGMENTS

Many thanks to my many colleagues at Mayo Clinic, the American Physiology Society, the Society of Women's Health Research, the Organization for the Study of Sex Differences, the International Society for Gender Medicine who have influenced and mentored me over the years. Our research studies would not have been possible without funding from the National Institutes of Health, the Kronos Longevity Research Institute, and the Mayo Foundation. Thanks to my daughter, Katherine E. Miller, PhD, for her support and helpful critique and editing of this chapter. May your career despite the challenges be as fulfilling as mine has been.

REFERENCES

[1] Bush TL, Barrett-Connor E, Cowan LD, Criqui MH, Wallace RB, Suchindran CM, et al. Cardiovascular mortality and noncontraceptive use of estrogen in women: results from the lipid research clinics program follow-up study. Circulation 1987;75:1102–9.

[2] Gisclard V, Miller VM, Vanhoutte PM. Effect of 17-β estradiol on endothelium-dependent responses in the rabbit. J Pharmacol Exper Ther 1988;244:19–22.

[3] Moraghan T, Antoniucci DM, Grenert JP, Sieck GC, Johnson C, Miller VM, et al. Differential response in cell proliferation to beta estradiol in coronary artery vascular smooth muscle cells obtained from mature female versus male animals. Endocrinology 1996;137:5174–7.

[4] Barber DA, Sieck GC, Fitzpatrick LA, Miller VM. Endothelin receptors are modulated in association with endogenous fluctuations in estrogen. Am J Physiol 1996;271:H1999–2006.

[5] Barber DA, Burnett Jr. JC, Fitzpatrick LA, Sieck GC, Miller VM. Gender and relaxation to C-type natriuretic peptide in porcine coronary arteries. J Cardiovasc Pharmacol 1998;32:5–11.

[6] Wang X, Barber DA, Lewis DA, McGregor CGA, Sieck GC, Fitzpatrick LA, et al. Gender and transcriptional regulation of endothelial nitric oxide synthase and endothelin-1 in porcine aortic endothelial cells. Am J Physiol 1998;273:H1962–7.

[7] Bracamonte MP, Rud KS, Owen WG, Miller VM. Ovariectomy increases mitogens and platelet-induced proliferation of arterial smooth muscle. Am J Physiol Heart Circ Physiol 2002;283:H853–60.

[8] Lewis DA, Bracamonte MP, Rud KS, Miller VM. Selected contribution: effects of sex and ovariectomy on responses to platelets in porcine femoral veins. J Appl Physiol 2001;91: 2823–30.

[9] Jayachandran M, Hayashi T, Sumi D, Thakur NK, Kano H, Ignarro LJ, et al. Up-regulation of endothelial nitric oxide synthase through b2-adrenergic receptor: The role of b-blocker with NO releasing action. Biochem Biophys Res Commun 2001;280:589–94.

[10] Jayachandran M, Miller VM. Age-dependent changes in functional factors of platelets from female humans. J Mol Cell Cardiol 2001;33:A52.

[11] Miller VM, Hay M. Principles of sex-based differences in physiology. Elsevier Publishing Company; 2004.

[12] Wizemann TM, Pardue ML. Exploring the biological contributions to human health: does sex matter? Board on health sciences policy. Washington, DC: Institute of Medicine; 2001.

[13] Langer RD. The evidence base for HRT: what can we believe? Climacteric 2017;20:91–6.

[14] Harman SM, Brinton EA, Cedars M, Lobo R, Manson JE, Merriam GR, et al. KEEPS: The Kronos Early Estrogen Prevention Study. Climacteric 2005;8:3–12.

[15] Miller VM, Black DM, Brinton EA, Budoff MJ, Cedars MI, Hodis HN, et al. Using basic science to design a clinical trial: baseline characteristics of women enrolled in the Kronos Early Estrogen Prevention Study (KEEPS). J Cardiovasc Transl Res 2009;2:228–39.

[16] Jayachandran M, Litwiller RD, Lahr BD, Bailey KR, Owen WG, Mulvagh SL, et al. Alterations in platelet function and cell-derived microvesicles in recently menopausal women: relationship to metabolic syndrome and atherogenic risk. J Cardiovasc Transl Res 2011;4:811–22.

[17] Miller VM, Petterson TM, Jeavons EN, Lnu AS, Rider DN, Heit JA, et al. Genetic polymorphisms associated carotid artery intima-media thickness and coronary artery calcification in women of the Kronos Early Estrogen Prevention Study. Physiol Genomics 2013;45:79–88.

[18] Farr JN, Khosla S, Miyabara Y, Miller VM, Kearns AE. Effects of estrogen with micronized progesterone on cortical and trabecular bone mass and microstructure in recently postmenopausal women. J Clin Endocrinol Metab 2013;98:E249–57.

[19] Raz L, Jayachandran M, Tosakulwong N, Lesnick TG, Wille SM, Murphy MC, et al. Thrombogenic microvesicles and white matter hyperintensities in postmenopausal women. Neurology 2013;80:911–18.

[20] Files JA, Miller VM, Cha SS, Pruthi S. Effects of different hormone therapies on breast pain in recently postmenopausal women: findings from the Mayo Clinic KEEPS Breast Pain Ancillary Study. J Womens Health 2014;23:801–5.

[21] Harman SM, Black DM, Naftolin F, Brinton EA, Budoff MJ, Cedars MI, et al. Arterial imaging outcomes and cardiovascular risk factors in recently menopausal women: a randomized trial. Ann Intern Med 2014;161:249–60.

[22] Gleason CE, Dowling NM, Wharton W, Manson JE, Miller VM, Atwood CS, et al. Effects of hormone therapy on cognition and mood in recently postmenopausal women: findings from the randomized, controlled KEEPS-Cognitive and Affective Study. PLoS Med 2015;12:e1001833.

[23] Miller VM, Lahr BD, Bailey KR, Heit JA, Harman SM, Jayachandran M. Longitudinal effects of menopausal hormone treatments on platelet characteristics and cell-derived microvesicles. Platelets 2015;27:1–11.

[24] Kling JM, Lahr B, Bailey K, Harman SM, Miller V, Mulvagh SL. Endothelial function in women of the Kronos Early Estrogen Prevention Study. Climacteric 2015;18:1–11.

[25] Miller VM, Jenkins GD, Biernacka JM, Heit JA, Huggins GS, Hodis HN, et al. Pharmacogenomics of estrogens on changes in carotid artery intima-medial thickness and coronary arterial calcification: Kronos Early Estrogen Prevention Study. Physiol Genomics 2016;48:33–41.

[26] Miller VM, Lahr BD, Bailey KR, Hodis HN, Mulvagh SL, Jayachandran M. Specific cell-derived microvesicles: linking endothelial function to carotid artery intima-media thickness in low cardiovascular risk menopausal women. Atherosclerosis 2016;246:21–8.

[27] Raz L, Hunter LV, Dowling NM, Wharton W, Gleason CE, Jayachandran M, et al. Differential effects of hormone therapy on serotonin, vascular function and mood in the KEEPS. Climacteric 2016;19:49–59.

[28] Kantarci K, Lowe VJ, Lesnick TG, Tosakulwong N, Bailey KR, Fields JA, et al. Early postmenopausal transdermal 17beta-estradiol therapy and amyloid-beta deposition. J Alzheimers Dis 2016;53:547–56.

[29] Kantarci K, Tosakulwong N, Lesnick TG, Zuk SM, Gunter JL, Gleason CE, et al. Effects of hormone therapy on brain structure: a randomized controlled trial. Neurology 2016;87: 887–96.

[30] Miller VM, Garovic VD, Kantarci K, Barnes JN, Jayachandran M, Mielke MM, et al. Sex-specific risk of cardiovascular disease and cognitive decline: pregnancy and menopause. Biol Sex Differ 2013;4:6.

[31] Miller VM. In pursuit of scientific excellence - sex matters. J Appl Physiol 2012;112:1427–8.

[32] Taylor KE, Vallejo-Giraldo C, Schaible NS, Zakeri R, Miller VM. Reporting of sex as a variable in cardiovascular studies using cultured cells. Biol Sex Differ 2011;2:11.

[33] Miller VM, Kaplan JR, Schork NJ, Ouyang P, Berga SL, Wenger NK, et al. Strategies and methods to study sex differences in cardiovascular structure and function: a guide for basic scientists. Biol Sex Differ 2011;2:14.

[34] Miller VM, Best PJ. Implications for reproductive medicine: sex differences in cardiovascular disease. Sex Reprod Menopause 2011;9:21–8.

[35] Miller VM, de Vries G, Arnold A. Sex and gender matters in drug development and life cycle management. Global Forum 2010;2:82–3.

[36] Chahal AA, Alhurani RE, Mohamed EA, Somers VK, Miller VM, Murad MH, et al. Are there sex differences following treatment of the left ventricular outflow tract obstruction in adults with hypertrophic cardiomyopathy? Eur Heart J Quality Care and Clinical Outcomes 2017;3:249–50.

[37] Schiebinger L, Leopold SS, Miller VM. Editorial policies for sex and gender analysis. The Lancet 2016;388:2841–2.

[38] Miller VM, Flynn PM, Lindor KD. Evaluating sex and gender competencies in the medical curriculum: a case study. Gend Med 2012;9:180–186.e3.

[39] Kling JM, Rose SH, Kransdorf LN, Viggiano TR, Miller VM. Evaluation of sex- and gender-based medicine training in post-graduate medical education: a cross-sectional survey study. Biol Sex Differ 2016;7:38.

[40] Miller VM, Kararigas G, Seeland U, Regitz-Zagrosek V, Kublickiene K, Einstein G, et al. Integrating topics of sex and gender into medical curricula-lessons from the international community. Biol Sex Differ 2016;7:6.

[41] Jenkins MR, Miller VM. 21st Century women's health: refining with precision. Mayo Clin Proc 2016;91:695–700.

CHAPTER 16

A Short History of the International Society for Gender Medicine (IGM)

Marek Glezerman[1,2] and Vera Regitz-Zagrosek[3]
[1]Rabin Medical Center, Research Center for Gender and Sex-Specific Medicine, Petah Tikva, Israel [2]Tel Aviv University, Sackler Medical School, Tel Aviv-Yafo, Israel [3]Institute for Gender in Medicine and Center for Cardiovascular Research, Charite, University Medicine Berlin DZHK, partner site Berlin

The first international congress on Gender Medicine was held in 2006 Berlin, before the birth of the International Society for Gender Medicine (IGM). In the same year, shortly after this event the IGM was officially founded in Berlin. As required by German law, there were seven founding members, five of whom served on the first board: Marianne Legato (United States), Maria Grazia Modena (Italy), Vera Regitz-Zagrosek (Germany), Karin Schenck-Gustafsson (Sweden), and Jeannette Strametz-Juranek (Austria). The new board agreed on the goals and structure of the society, i.e., to promote gender medicine in research, teaching, and patient management. Vera Regitz-Zagrosek served as founding president.

The initial structure included a rotation of presidency among the board members that organized the annual or bi-annual congresses. The 2nd congress took place in Vienna in 2007 and the 3rd congress was held in 2008 in Stockholm, under the presidency of Karin Schenck-Gustafsson. At this meeting, Marek Glezerman from Israel joined the board. The 4th meeting took place again in Berlin in 2009 and Vera Regitz-Zagrosek presided. For organizational and financial reasons, the meeting was held in collaboration with the German Society of Cardiothoracic Surgeons, led by Roland Hetzer. Science as well as the structure of IGM were discussed. The general assembly re-elected the acting board members.

The 5th International Congress of Gender- and Sex-Specific Medicine (GSSM) took place in Tel Aviv with the participation of over 450 professionals from 18 countries. During this meeting, Marek

The International Society for Gender Medicine. DOI: http://dx.doi.org/10.1016/B978-0-12-811850-4.00016-8

Glezerman was elected president of IGM. The board defined its priorities: We needed urgently to try and coordinate between the many different groups and organizations which were interested in GSSM, we needed to increase our membership which at that time totaled worldwide about 120 professionals, we needed working bylaws, and we needed to attract other national and professional societies to join the IGM. We also needed continuity in organizing international meetings and we needed to establish a closer relationship with the American Organization for the Study of Sex Differences (OSSD).

Have we succeeded so far? To a certain extent, yes. The 6th International Congress was organized jointly by the OSSD and IGM in Baltimore, mostly due to the efforts of Prof. Virginia Miller from the Mayo Clinics in Rochester, who later was also elected to the IGM board. The meeting was very successful, but unfortunately, subsequent joint meetings did not follow. Funding of the IGM became a real issue. Being dependent on membership fees which were very modest, it became increasingly difficult to make ends meet. Fortunately, the Foundation for Gender Medicine in New York, founded by Marianne Legato came to the rescue with a very generous donation which permitted us to move on. The activities of all board members were always completely voluntary and no one ever even asked for disbursement of expenses.

The next great step followed a small lecture series which Marek gave in Japan in 2013. It was there where he met Prof. Hiroaki Shimokawa, the president of the Japanese Society for Gender Medicine and also one of the leading cardiologists in his country. Shortly afterward we were very happy to welcome the Japanese Society for Gender medicine as full member to the IGM and Hiroaki was eventually elected to the board.

The last group to join the IGM was Eugennet, an international group of professionals who were involved in a multicenter program, sponsored by the European Union and led by Vera Regitz-Zagrosek. As of this writing, the IGM has close to 750 members, including member societies in Austria (presided by Prof. Alexandra Kautzky-Willer), Germany (presided by Prof. Vera Regitz Zagrosek), Israel (presided by Prof. Marek Glezerman), Italy (presided by Prof. Giovanella Baggio), Italy (presided by Prof. Elvira Reale), Japan (presided by Prof. Prof. Hiroaki Shimokawa, Sweden (presided by Prof. Schenck-Gustafsson), United States (presided by Prof. Marianne Legato), and individual members from Brazil, India, Canada, Russia, Mexico, and the Netherlands.

As time passed, the board underwent additional changes: Prof. Giovanella Baggio, a very well-known Italian internist and president of the Italian Society for Gender medicine joined the board, Prof. Jeanette Strametz-Juranek resigned for personal reasons, and Prof. Alexandra Kautzky-Willer, a renowned endocrinologist and current president of the Austrian Society for GSSM joined. The 7th International meeting was coming up and with it our exposure to the tremendous difficulties in organizing these meetings: Funding from industry had become more and more rare and the amounts invested smaller and smaller. What is more important, industry is not overly interested in funding multidisciplinary events. This is understandable, since GSSM is by definition a heterogeneous discipline, actually encompassing all fields of medicine. This appears to be a disadvantage for funding by pharmaceutical companies. If they are interested in marketing a specific drug, they will find it difficult to identify the target population at an ISSM meeting and will prefer to invest in a scientific meeting with a more discrete focus. The organization of the 7th meeting was a nightmare. Initially, Japan and Italy assessed the possibility of organizing the meeting, but Prof. Maria Grazia Modena realized soon that there would not be sufficient funding available. The Japanese option also became obsolete because a possible collision with another major meeting which Prof. Shimokawa was organizing. The Israel Society then decided to take up the challenge. Contracts were signed with an organizing company, the venue for the meeting was secured, invitations were sent out, a dedicated website went online, and quite some money was spent up front. But eventually, severe funding restraints forced us to give up. All of us were deeply frustrated until Prof. Vera Regitz-Zagrosek came to the rescue. Facing the same difficulties as all others who had failed, Vera made the impossible possible and organized a superb meeting in Berlin. This was also made possible by the enormous input of all board members and scientists and furthermore by a synergistic work with a German congress on Gender medicine, namely "Junior meets senior," sponsored by the Federal German Ministry of Education and Research (BMBF). More than 300 attendees came together and the open discussions in different formats was appreciated by all.

Yet, given the mounting difficulties in organizing a congress, who would dare to take upon herself/himself to organize the following meeting, the Eighth International congress? With great relief and excitement everybody on the board welcomed the offer by Prof. Shimokawa who volunteered to take up the challenge and to host the next meeting in Japan. At this writing, we are all looking forward to September 2017, knowing extremely well what an effort is behind such a project and how much effort goes into the struggle for funding from meeting to meeting.

The other, seemingly easy issue was to put together working bylaws, indeed a much larger endeavor than one would have anticipated. Marek came to realize this very quickly after having taken up the task. Today, our bylaws are based on more than a dozen bylaws of similar professional organizations and the very valuable input of our board members. The blueprint was scrutinized by legal counsel and we are all proud that our activities are based today on solid legal grounds and worthy of a society like our IGM. The bylaws deal with day to day business of the society, with election procedures and with rights and duties of members and officers. But above all, the bylaws define the framework within which the society is active and the goals for which it exists. We have stated in our bylaws that the specific purpose of our society would be to establish and develop gender medicine in an international context by promoting gender-specific research in basic sciences, clinical medicine, and public health. Specifically, our society would aim to

- advance the understanding of sex/gender differences by bringing together scientists and clinicians of diverse backgrounds;
- strive to implement gender in the medical curriculum, prepare and allocate gender-specific learning materials, curricula and gender trainings for instructors;
- promote gender-specific public health issues such as information for persons, institutions, and organizations in the area of gender medicine;
- facilitate interdisciplinary research on sex/gender differences in basic and clinical frameworks;
- encourage the application of new knowledge of sex/gender differences to improve health and health care;
- cooperate with other professional or national and international societies of gender medicine and similar scientific organizations;
- encourage and support the creation of professional or national organizations dedicated to the promotion of sex/gender medicine;

- encourage and support international cooperation, collaboration, and education among professionals working in the field of sex/ gender medicine;
- organize international meetings and congresses on relevant topics.

What have we accomplished so far related to these goals? We dare to say, quite a lot, although there is still much more ahead to be done.

We started off in 2009 with approximately 120 members worldwide and we have at this writing almost 750 members in 14 countries. As you can read in the contributions of the authors of this book, gender and sex specific medicine has been integrated into the curricula of medical studies in many leading universities. Summer schools and post-graduate courses are being offered, fellowships are available and despite all hardships, scientific meetings are being held worldwide, PhD programs are being conducted and research is mushrooming in basic sciences, in clinical fields, and in translational research. Large multinational studies, such as Eugenmed, are being conducted, sponsored by regulators. The number of scientific publications related to GSSM is mushrooming [1–3], excellent textbooks have been published [4–9], books for the general public are being offered in increasing numbers [10–12] and dedicated journals to GSSM are available [13–14]. There are still many fundamental issues which IGM as a society and certainly the member societies need to address: For example, should GSSM aim to become a separate and distinctive medical discipline, should the understanding of GSSM become an integral part of every medical discipline or are these two options actually not a contradiction and we should strive to achieve both? And if, so, what should be the framework? What should be the interaction between personalized medicine and GSSM? Is GSSM just a step in the direction of personalized medicine or do the two complement each other? And if so, how? And a seemingly merely semantic question: Even throughout this book, you will find many different names for our new science. Even the name of our society is actually a misnomer, since we are not solely dealing with the sociological gender construct in medicine but also with the biological category, namely "sex." Is it not about time that we all at least agree on a common term? Hopefully, at the time when you, dear reader, are holding this text in your hand, at least IGM will have changed its name already, and hopefully other organizations will follow course.

We are proud that the IGM and its member societies have contributed so much to the promotion of GSSM and to raising the awareness amongst regulators, academic institutions, and our patients. The media has become very much aware of the new science of GSM; this needs to be promoted even more. The more we involve the media, the more we reach out to the general public, the stronger becomes the grassroots movement of our patients who address their physicians with questions related to gender aspects of health and disease. And these questions can turn out to be great motivators for opening the minds of practioners. After a little more than a decade, International GSSM has come a long way, but we are still at the beginning. Concerted efforts are required involving professionals and lay persons, legislators and regulators, academia and service providers, teachers and students, clinicians and patients—all of us need to continue and promote this new discipline as a major step in the improvement of medical care.

REFERENCES

[1] Regitz-Zagrosek V, Kararigas G. Mechanistic Pathways of sex differences in cardiovascular disease. Physiol Rev 2017;97:1–37.

[2] EUGenMed Cardiovascular Clinical Study Group, Regitz-Zagrosek V, Oertelt-Prigione S, Prescott E, Franconi F, Gerdts E, et al. Gender in cardiovascular diseases: impact on clinical manifestations, management, and outcomes. Eur Heart J 2016;37:24–34.

[3] McGregor A, Choo EK, Becker BM. Sex and Gender in acute care medicine. New York: Cambridge University Press; 2016.

[4] Gender Medizin. In: Rieder A, Lohff B, editors. Geschlechtsspezifische Aspekte für die klinische Praxis. 2nd ed. Wien, New York: Springer; 2010.

[5] Schenck-Gustafsson K, editor. Handbook of clinical gender medicine. Basel: Karger; 2012.

[6] Kautzy-Willer A. Gendermedizin. Wien: Boehlau Verlag; 2012.

[7] Oertelt-Prigione S, Regitz-Zagrosek V, editors. Sex and gender aspects of clinical medicine. London: Springer Verlag; 2012.

[8] Regitz-Zagrosek V. Sex and gender differences in pharmacology. Berlin, Heidelberg: Springer Verlag; 2014.

[9] Legato MJ, editor. Principles of gender-specific medicine. 3rd ed. San Diego: Elsevier; 2017.

[10] Legato JM. Eve's Rib. New York: Harmony Books; 2002.

[11] Legato JM. Why men die first. New York and Houndmills: Palgrave MacMillan; 2008.

[12] Glezerman M. Gender medicine. New York and London: Overlook/Duckworth; 2016.

[13] Biology of sex differences. London: BioMed Central.

[14] Gender and the Genome. New Rochelle: Mary Ann Liebert, Inc.

INDEX

A

AIDS, 96–98
AIFA. *See* Italian Pharma Agency (AIFA)
Allgemeiner Kranken Haus (AKH), 111
American Heart Association, 2–3, 136–137
American Medical Women's Association, 123–124, 126–127
AMMI. *See* Associazione Mogli e Mariti di Medici Italiani (AMMI)
Appalachia to West Texas
  Appalachian coal country, 117
  Chair of Internal Medicine, 119
  First Lady of United States, 122
  gender-health trajectory, 123
  Institute of Medicine, 120
  Midland Texas, 121
  personalized medicine, 124
  sex and gender medicine, 124
  SGBM, 118
  TTUHSC, 119–120
Associazione Mogli e Mariti di Medici Italiani (AMMI), 66–67
Austria, Gender Medicine in, 23–24, 27. *See also* Italy—GSM in
  diabetes, 23–24, 28
  EASD, 24
  endocrinology and metabolism, 26
  gender dimorphism, 26–27
  interdisciplinary science, 25
  International Society of Gender Medicine, 24–25
  obesity-related complications, 28
  "-omics", 29
  one-size-fits-all medicine, 30
  Possanner von Ehrenthal Award, 31
  sex-specific aspects, 25, 30
  type 1 diabetes, 26
Austrian Society of Gender Specific Medicine (ÖGGSM), 24–25, 37–38
Autoimmune diseases, 26

B

Bikini Medicine, 17
Biology, 1, 110
  gender-specific, 3–4
  molecular, 20
  reproductive, 45
Biomedical research, 17–21, 30, 124
BMBF. *See* Federal German Ministry of Education and Research (BMBF)
Board members, 143–147
Boston Women's Health Collective, 45–46
Bylaws, 146–147

C

CAD. *See* Coronary artery disease (CAD)
Canadian Institutes of Health Research, 51
Cardiometabolic diseases, 24–25
Cardiovascular Disease in women, 58
Cardiovascular diseases (CVD), 33, 80, 107–108, 111–114
  future, gender medicine in, 39–42
  Gender Mainstreaming and Diversity at Medical University of Vienna, 34–35
  gender-sensitive research and teaching, 35–36
  MI, 33
  MVD, 34
  past, gender medicine in, 36–38
    Austrian Society of Gender Specific Medicine, 37–38
    education and research, 36–37
  present, gender medicine in, 38–39
  underlying diseases and referral to rehabilitation 2016, 39
Cardiovascular research, 57–59
CCU. *See* Coronary care unit (CCU)
Center for Gender Medicine (CfG), 109, 111–112
  establishment, 14
  Karolinska Institutet
    cardiac rhythm disturbances, 108
    creating center for research and education, 110–112
    CVD, 107–108
    education, 110–112
    left ventricular function, 108
    Men's health, 109
    national and international activities, 112–113
    research, 113–114

Center for Gender Medicine (CfG) (*Continued*)
at Rabin Medical Center, 13–14
Center for Gender Specific Medicine at Columbia University, 113
Center of National Health and Gender Medicine, 69
Centro Studi Italiano per la Salute e la Medicina di Genere (CSSMG), 66–69
CfG. *See* Center for Gender Medicine (CfG)
Chair of Gender Medicine at University of Padua, 75–76
Charite approach, 57
gender in research, 59–60
from student to professor in gender medicine, 57–59
teaching, 60–62
Children, 117
Clinical research, 20
CME. *See* Continuing medical education (CME)
Collaborative Specialization in Women's Health, 49–50
Columbia, 3–5
"*Come cambia la vita delle donne*", 85–86
Continuing medical education (CME), 72–73
Coronary artery disease (CAD), 2–3, 81
Coronary care unit (CCU), 107
CSSMG. *See Centro Studi Italiano per la Salute e la Medicina di Genere* (CSSMG)
CVD. *See* Cardiovascular diseases (CVD)

D

Depression, 28, 96–100, 109
DGesGM. *See* German Society of Gender Medicine (DGesGM)
Diabetes, 23–24, 28
gestational, 23–24
type 1, 23–24, 26
type 2, 23–24
Disability, 40
Drug treatment, 87–90

E

EASD. *See* European Association for the Study of Diabetes (EASD)
ED. *See* Emergency department (ED)
EDRF. *See* Endothelium-derived relaxing factor (EDRF)
Education, 27–28, 31
EM. *See* Emergency Medicine (EM)
Emergency department (ED), 125, 127–128, 130
Emergency Medicine (EM), 125
sex and gender in, 125–131
Endocrinology, 26
Endothelial dysfunction, 101–103
Endothelium, 134–135
Endothelium-derived relaxing factor (EDRF), 134–135
Epistemology, 51–52
Equity Act, 86–87
ESC. *See* European Society of Cardiology (ESC)
Estrogen, 26, 59, 134–137
EUGENMED. *See* European Gender Medicine (EUGENMED)
EUGIM. *See* European curriculum in Gender Medicine (EUGIM)
European Association for the Study of Diabetes (EASD), 24
European curriculum in Gender Medicine (EUGIM), 59–60
European Gender Medicine (EUGENMED), 59–60
European Medical Agency, 101
European Society of Cardiology (ESC), 59–60, 101–103

F

FADOI. *See* Italian Federation of Hospital Internal Medicine Specialists (FADOI)
FDA. *See* Food and Drug Administration (FDA)
Federal German Ministry of Education and Research (BMBF), 145
Federation of all Italian General Medical Council (FNOMCeO), 67–68, 72–73
Federazione Italiana dell Società Mediche (FISM), 74
FNOMCeO. *See* Federation of all Italian General Medical Council (FNOMCeO)
Food and Drug Administration (FDA), 101, 113
Functioning, 40
Funding, 144–145
NIH-funded studies, 137

G

GENCAD. *See* Gender in coronary artery disease (GENCAD)
Gender, 86–87, 117
Button, 113–114
challenge, 71
dimorphism, 26–27
lectures, 35–36

pharmacology, 87–90
in research, 59–60
research, 90
Gender in coronary artery disease (GENCAD), 59–60
Gender Mainstreaming and Diversity at Medical University of Vienna, 34–35
Gender medicine (GM), 65–66, 95, 143. *See also* Gender-specific medicine (GSM); International Society for Gender Medicine (IGM)
curriculum in, 62
in Italy, 95
barriers, 101
gender, 98
last, personal, barrier, 101–104
limits and plans for future, 104–105
occupational health, 99–100
Report on Women's Health, 96–98
RFs, 98–99
specific aims, 96
qualification in, 62
from student to professor in, 57–59
Gender-and Sex-Specific Medicine (GSSM), 10–14, 143–144, 147–148
Gender-based medicine, 30, 35
Gender-health trajectory, 123
Gender-sensitive research and teaching, 35–36
gender lectures, 35–36
postgraduate program gender medicine, 36
Gender-specific medicine (GSM), 1, 5–13, 17, 34, 46, 80–81. *See also* Gender medicine (GM)
crucial role of, 82
Menarini International Foundation, 79
nonprofit educational foundation, 81
physiologic and pathological mechanisms, 80
prediabetic women, 81
programs to international level, 79–80
in scientific programs, 82
gendeRingvorlesungen, 36–37
"Genome and Hormones: An Integrated Approach to Gender Differences in Physiology", 136
German law, 143
German Society of Gender Medicine (DGesGM), 62
Gestational diabetes, 23–24
GiM. *See* Institute of Gender in Medicine (GiM)
Giovanni Lorenzini Medical Science Foundation in Gender Medicine, 71
GISeG. *See* Italian Group for Health and Gender (GISeG)
GM. *See* Gender medicine (GM)
GSM. *See* Gender-specific medicine (GSM)
GSSM. *See* Gender-and Sex-Specific Medicine (GSSM)

H

Handbook of Clinical Gender Medicine, 111–112
Health, sex and gender in
call for action, 52
GSM, 46
journey, 46–48
"Our Bodies, Ourselves", 45–46
struggles and support, 50–52
world writes on body, 48–49
Heart
diseases, 81
failure, 108
Hormonal replacement therapy (HRT), 95, 101–103
Hormones and Behavior, 46–47
HRT. *See* Hormonal replacement therapy (HRT)
Human genome, 20–21
Human medicine, 35
Hypertension, 96–98

I

ICD. *See* International Classification of Diseases (ICD)
ICF. *See* International Classification of Functional Disability Criteria (ICF)
IGH. *See* Institute of Gender and Health (IGH)
IGM. *See* International Society for Gender Medicine (IGM)
Institute of Gender and Health (IGH), 51
Institute of Gender in Medicine (GiM), 58–59, 62
Institute of Medicine, 120, 136
Interdisciplinary science, 25
International Classification of Diseases (ICD), 40–41
International Classification of Functional Disability Criteria (ICF), 38–41
International Network of Health Promoting Hospitals and Health Services, 38–39
International Society for Gender Medicine (IGM), 37–38, 65–66, 112–113, 137–138, 143. *See also* Gender medicine (GM)

International Society for Gender Medicine (IGM) (*Continued*)
bylaws, 146–147
GSSM, 145
5th International Congress, 143–144
International, 148
international group of professionals, 144–145
scientific publications, 147
ISC. *See* Italian Society of Cardiology (ISC)
Israel Society for Gender Medicine (IsraGem), 14
GSSM, 10, 13–14
reproductive endocrinology, 9–10
IsraGem. *See* Israel Society for Gender Medicine (IsraGem)
ISS. *See Istituto Superiore di Sanita* (ISS)
Istituto Nazionale di Statistica, 85–86
Istituto Superiore di Sanita (ISS), 69
Italian Federation of Hospital Internal Medicine Specialists (FADOI), 67–68, 72–74
Italian Group for Health and Gender (GISeG), 70–73
Italian Pharma Agency (AIFA), 101
Italian Society of Cardiology (ISC), 101–103
Italy
GM in, 95. *See also* Austria, Gender Medicine in
barriers, 101
gender, 98
last, personal, barrier, 101–104
limits and plans for future, 104–105
occupational health, 99–100
Report on Women's Health, 96–98
RFs, 98–99
specific aims, 96
GSM in
Chair of Gender Medicine at University of Padua, 75–76
communication tools within and outside network, 76–77
educational activities, 66
information activities, 66–67
Italian Network, 67–68
Italian universities, 75
Lorenzini Foundation, 65
network creation, 67–75
research, 67
Italian College of General Practitioners, 74
Italian Constitution, 85
Italian Federation of Medical Councils, 72
Italian general contest, 85–86
Italian Journal of Gender-Specific Medicine, 76–77
Italian Journal of Medicine, 73
Italian National Medical Council, 72–73
Italian Political Dimension, 77
Italian universities, 75

J

Journey into sex and gender medicine field, 133–134
comparative and integrated physiology, 138–139
creative approaches, 134
funding mechanisms to NIH, 138
progress in spite of rocky roads, 136–138
trail blazers, 135–136

K

Karolinska Institutet (KI), 86–87, 109
Center for Gender Medicine
cardiac rhythm disturbances, 108
creating center for research and education, 110–112
CVD, 107–108
education, 110–112
left ventricular function, 108
Men's health, 109
national and international activities, 112–113
research, 113–114
KEEPS. *See* Kronos Early Estrogen Prevention Study (KEEPS)
KI. *See* Karolinska Institutet (KI)
Knowledge, 17–18, 21
basis, 61
KOK. *See* Stockholm Female Coronary Artery Study (KOK)
Kronos Early Estrogen Prevention Study (KEEPS), 136–137

L

Left ventricular function, 108
Left ventricular systolic dysfunction, 108
Lega Italiana contro I Tumori (LILT), 66–67
Lorenzini Foundation, 65, 67–68

M

"Male norm", 133–134
Master Plan Rehabilitation, 39–42
Medical care, 17
Medical contest, 86–87
Medical Performance Profiles (MLP), 40

Medical practice, 82
Medical University of Vienna, 24–25, 27–28
MEGE. *See* National Center for Gender-Specific Medicine (MEGE)
Memberships, 1
Memorandum of Understanding (MOU), 113
Menarini International Foundation, 79
Menopause, 101–103, 122
Metabolism, 26
MI. *See* Myocardial infarction (MI)
Microvascular disease (MVD), 33–34
Midland Texas, 121
Ministry of Health, 104–105
MLP. *See* Medical Performance Profiles (MLP)
MOU. *See* Memorandum of Understanding (MOU)
MVD. *See* Microvascular disease (MVD)
Myocardial hypertrophy, 58
Myocardial infarction (MI), 33, 107

N

National and international activities, CfG KI, 112–113
National Center for Gender-Specific Medicine (MEGE), 69–70
  of ISS, 69, 76
National Institutes of Health (NIH), 18, 112–113, 135
National Research Center for Gender Health and Medicine. *See Centro Studi Italiano per la Salute e la Medicina di Genere* (CSSMG)
National Statistical Institute, 96–98
New Karolinska Hospital (NKS), 107
Newsletter, 76
  Italian Journal of Gender-Specific Medicine, 76–77
NIH. *See* National Institutes of Health (NIH)
NIH Office of Research on Women's Health, 136
NIH Revitalization Act (1993), 18
NIH-funded studies, 137
Nitric oxide (NO), 134–135
NKS. *See* New Karolinska Hospital (NKS)
NO. *See* Nitric oxide (NO)
Non-Caucasian subjects, 89
Normal science, 50

O

Obesity-related complications, 28
Obstetrics and gynecology, 9–13
Occupational health, 99–100
Office of Research on Women's Health (ORWH), 18, 135
ÖGGSM. *See* Austrian Society of Gender Specific Medicine (ÖGGSM)
"-omics", 29
One-size-fits-all medicine, 30
Operating Units, 69–70
Organization for Study of Sex Differences (OSSD), 9–10, 112–113, 137–138, 143–144
ORWH. *See* Office of Research on Women's Health (ORWH)
OSSD. *See* Organization for Study of Sex Differences (OSSD)

P

P&G, 3–5
Pathophysiology, 10–13
Pensionsversicherungsanstalt, 39
Personalized medicine, 124
Pharmacodynamic differences, 87–88
Pharmacology, 90
Possanner von Ehrenthal Award, 31
Postgraduate program gender medicine, 36
Precision medicine, 21, 28–30
  Initiative, 20–21
Prediabetic women, 81
Principles of Gender-Specific Medicine, 5, 120, 126
Psychiatry, 13–14, 60–61, 98–99, 107, 109
Psychosocial risk factors, 108

R

Rabin Medical Center, 13–14
Rehabilitation program, 38–39
Reproductive endocrinology, 9–10
Reproductive medicine, 127–128
Risk factors (RFs), 95, 98–99, 101–103

S

SAEM. *See* Society for Academic Emergency Medicine (SAEM)
Schizophrenia, 96–98
Science, 50
Sex, 86–87
  differences, 46–48, 51–52
  sex-specific effects, 30
Sex and gender
  aspect, 57
  in Cardiovascular Research, 111
  differences, 24–25
  in emergency medicine, 125–131

Sex and gender (*Continued*)
Institute of Medicine, 120
medicine, 124
sex-and gender-based factors, 128
Sex and Gender in Emergency Medicine (SGEM), 128–129, 131
Community Advisory Board, 128–129
Sex and Gender Women's Health Collaborative (SGWHC), 126–127
Sex and the Brain, 46–47
Sex-and gender-based medicine (SGBM), 117–118, 138
sex-and gender-specific medicine, 133–134
in United States
Bikini Medicine, 17
biomedical research, 17–18
clinical research, 20
human genome, 20–21
NIH, 18
NIH ORWH, 19–20
precision medicine, 21
Women's health priorities, 20
Sexual dimorphisms, 47
SGBM. *See* Sex-and gender-based medicine (SGBM)
SGEM. *See* Sex and Gender in Emergency Medicine (SGEM)
SGWHC. *See* Sex and Gender Women's Health Collaborative (SGWHC)
SIMG. *See* Society of General Medicine Doctors (SIMG)
SKA. *See* Sonderkrankenanstalt (SKA)
Skepticism, 10–13
overcoming
gender pharmacology, 87–90
Italian general contest, 85–86
medical contest, 86–87
pharmacological response beyond drug factors, 87–88
Society for Academic Emergency Medicine (SAEM), 126
Society for Women's Health Research (SWHR), 52
Society of General Medicine Doctors (SIMG), 67–68
Sonderkrankenanstalt (SKA), 38
Sonderkrankenanstalt Rehabilitationszentrum Bad Tatzmannsdorf, 37
Steroid hormones, 26
Stockholm County Council, 110–111
Stockholm Female Coronary Artery Study (KOK), 108
Stress, 98–100
stress-related diseases, 96–98
SWHR. *See* Society for Women's Health Research (SWHR)

T

Teaching, 60–62
book, 62
curriculum in gender medicine, 62
medical students, 60–62
nonmedical students and professionals, 61–62
qualification in gender medicine, 62
Testosterone, 26
Texas Tech University Health Sciences Center (TTUHSC), 119–121
"The Gender Lens", 110–111
The Principles of Gender-Specific Medicine, 120
Tobacco smoking, 98
Torsade de pointe-ventricular tachycardia, 108
Trail blazers, 135–136
TTUHSC. *See* Texas Tech University Health Sciences Center (TTUHSC)
Type 1 diabetes, 23–24, 26
Type 2 diabetes, 23–24

U

US Food and Drug Administration's Office of Women's Health, 124

V

Vascular endothelium, 134–135
Violence, 96–100

W

Wallberg-Henriksson, Harriet, 109
Wegweiser durch die geschlechtsspezifische Rehabilitation, 39
West of Scotland Coronary Prevention Study (WOSCOPS), 89
WHEC. *See* Women's Health in Emergency Care (WHEC)
WHI. *See* Women's Health Initiative (WHI)
WHO. *See* World Health Organization (WHO)
WIC. *See* Women in Cardiology Committee (WIC)
Women in Cardiology Committee (WIC), 101–103

Women-oriented healthcare system, 96
"Women's Clinic", 10–13, 101–103
Women's Health, 48, 125–128, 138
priorities, 20
Women's Health Equity Act, 9–10
Women's Health in Emergency Care (WHEC), 127–128
Women's Health Initiative (WHI), 135–137
Committee, 126–127
World Health Organization (WHO), 86–87
WOSCOPS. *See* West of Scotland Coronary Prevention Study (WOSCOPS)

Printed in the United States
By Bookmasters